REDUCTION, TIME AND REALITY

Reduction, time and reality

Studies in the philosophy of
the natural sciences

Edited by Richard Healey

Assistant Professor of Philosophy
University of California at Los Angeles

CAMBRIDGE UNIVERSITY PRESS

CAMBRIDGE

LONDON   NEW YORK   NEW ROCHELLE
MELBOURNE   SYDNEY

CAMBRIDGE UNIVERSITY PRESS
Cambridge, New York, Melbourne, Madrid, Cape Town, Singapore,
São Paulo, Delhi, Dubai, Tokyo, Mexico City

Cambridge University Press
The Edinburgh Building, Cambridge CB2 8RU, UK

Published in the United States of America by Cambridge University Press, New York

www.cambridge.org
Information on this title: www.cambridge.org/9780521143721

© Cambridge University Press 1981

First published 1981
First paperback printing 2010

A catalogue record for this publication is available from the British Library

ISBN 978-0-521-23708-6 Hardback
ISBN 978-0-521-14372-1 Paperback

Contents

Contributors

Michael Friedman is Associate Professor of Philosophy at the University of Pennsylvania.

Peter Alexander is Professor of Philosophy at the University of Bristol.

D. C. Dennett is Professor of Philosophy at Tufts University.

P. C. W. Davies is Professor of Theoretical Physics at the University of Newcastle upon Tyne.

D. H. Mellor is Lecturer in Philosophy at Cambridge University, and Fellow of Darwin College.

Richard Healey is Assistant Professor of Philosophy at the University of California, Los Angeles.

Lawrence Sklar is Professor of Philosophy at the University of Michigan.

Colin McGinn is Lecturer in Philosophy at University College, London.

Bas C. van Fraassen is Professor of Philosophy at the University of Toronto and the University of Southern California.

Introduction

For much of the past fifty years, logical positivism constituted the dominant tradition in the philosophy of science. Positivists attempted to impose restrictions on the content of scientific theories in order to ensure that they were empirically meaningful. An effect of these restrictions was to limit both the claims to truth of theoretical sentences only distantly related to observation, and the claims to existence of unobservable theoretical entities. More recently positivism has come under such sustained attack that opposition to it has become almost orthodoxy in the philosophy of science. But a purely negative view cannot long command a consensus, and so following the demise of positivism the debate in the philosophy of science has shifted to its would-be successor, scientific realism.

In contrast to positivism, the key tenets of scientific realism may be expressed as follows: terms in a mature science typically refer to real objects; and the laws of a theory belonging to a mature science are typically approximately true. Much recent philosophy of science consists of arguments for and against scientific realism as so understood. A particularly influential type of objection to scientific realism appeals to the history of science to demonstrate its supposed inadequacy: the claim is that there is no way of telling to which objects, if any, particular theories referred, and no useful sense in which these theories were even approximately true; nor is there any reason to expect our current theories to prove different in these respects. A realist response is to claim that, on the contrary, only scientific realism renders intelligible the success of science.

The papers in this volume are concerned less with this well-aired controversy than with other, more specific issues concerning realism in science. Their general tendency is to question the terms in which the debate about scientific realism is usually conducted. Specifically, standard formulations of scientific realism fail to do justice to the diversity of issues which come under the general heading of realism in science. Our authors therefore examine on their own merits a number of different questions concerning the existence of various kinds of scientific entity, and the truth of various kinds of scientific statement. By this means, we hope that the volume as a whole will reveal a richness of texture in scientific and philosophic thought which is sometimes overlooked in the contemporary debate concerning scientific realism.

The articles cover three broad topics: reduction, time and modality. Reduction is the name given to that process by which one theory or branch of science comes to be incorporated into another theory or branch of science. A central question here is how a realist should regard scientific reductions. One simple answer would be that a realist should favour those reductions that show which entities in the reducing theory are identical to, or compose, the entities of the reduced theory, and how the approximate truth of the reducing theory accounts for that of the reduced. But, as the papers of Friedman, Alexander and Dennett illustrate, this answer is too simple: the scientific realist should only favour reductions in certain circumstances; awkward conceptual problems may face particular types of proposed reduction; and there may be cases in which a realist should reject the demand that entities in reduced and reducing theories be identified.

In the first paper of this volume, Michael Friedman addresses the general question: when does a theoretical reduction license a realistic attitude to the entities of the reducing theory? He proposes an answer in terms of the unifying power of the resulting theoretical structure, and uses this answer to provide a criterion for the desirability of particular reductions. Application of this criterion has, he argues, consequences unfavourable to possible world semantics and to the 'language of thought' hypothesis proposed by Fodor.

Peter Alexander examines a number of conceptual problems which he takes a certain type of reductive explanation to face. These problems concern the meaning rather than the reference of terms in the reducing theory. He pursues his examination in the context of a particular explanatory proposal, the atomism of Locke and Boyle, in which the problem is to explain how possession of each primary quality may intelligibly be attributed to a single corpuscle. For example, it may be thought meaningless to ascribe size to a single corpuscle in an otherwise empty universe. Alexander concludes, albeit tentatively, that despite the problems he discusses, such attributions can be made intelligible.

Psychology has long presented a special challenge to the reductionist. Can psychology be reduced to something else, and if so how, and to what? D. C. Dennett takes what he calls 'folk psychology' as the starting point of his answer to this question. He argues that though folk psychology is in many ways confused, it is primarily an instrumentalist calculus of beliefs and desires used to predict what people will do if they are more or less

Introduction

metaphysics of time. My paper is concerned with the direction of time, and with the relevance to it of a theory of contemporary physics, namely quantum mechanics. I offer two suggestions as to what might be meant by the assertion that time is objectively directed, and then consider attempts to argue that quantum mechanics shows time to be directed in one of these senses. Quantum mechanics is a statistical theory, and this proves very relevant to these arguments. I contend that quantum mechanics provides no good reason to think that time is objectively directed. The paper concludes with a suggestion as to how physical theory might more appropriately be used to show the objectivity of time's direction.

Lawrence Sklar investigates the bearing of another physical theory, the theory of relativity, on some of the metaphysical views mentioned above: namely, the view that only what exists now really exists, and the view that only the present and past have determinate existence. These views appear to run into difficulties when faced with the relativity of simultaneity (and hence of the present) inherent in relativity theory. Sklar argues that relativity does not refute such views, but that their most natural relativistic reformulations involve the denial of reality to the spatially distant as well as to the temporally distant. His paper also raises some deep problems for both proponents and opponents of realism.

Modal notions such as those of necessity and possibility are constantly employed by scientists, while philosophers have either questioned their legitimacy or disagreed about their correct analysis. The question as to whether there is real necessity or possibility is an important topic for the philosophy of science. But there is some difficulty in characterizing a realist view of these modalities, since no theory in the natural sciences refers to specifically modal objects as such. One may indeed interpret scientists' talk of necessity and possibility as implicitly referring to specifically modal objects (such as possible worlds). But a scientific realist need not be committed to these things, provided he can give an alternative analysis of such talk. And this raises two questions: how to give an adequate characterization of modal realism, and whether a scientific realist is committed to modal realism (and if so of what kind).

Colin McGinn's paper is devoted to a comprehensive examination of modal realism. He distinguishes objectual modal realism, which is committed to the existence of modal objects (possible worlds), from other forms of modal realism, which are not. In the first part of his paper argues against objectual modal realism in a way which may usefully compared to Friedman's discussion of this topic. In providing a char ization of non-objectual modal realism McGinn gives a more formulation of realism which contrasts both with scientific re formulated above, and also with Michael Dummett's influent

x

rational. Reduction is to be sought in two directions: first, by rational folk psychology into a coherent theory of intentional systems, analogo decision theory; second, by creating a theory of sub-personal cogn: psychology which explains how a part of the human organism like brain is able to extract meaningful information from its inputs. Neurop siology will enter only at the stage of explaining how intentiona characterized sub-personal operations and states are realized.

The next four papers concern the science and metaphysics of tim Contemporary science tells a detailed story about time. One might thin that, for the scientific realist, time is real just in case it is suitably referre to by our best scientific theories. But long-standing metaphysical dispute about time and reality seem not to be so easily resolved. McTaggart, for example, notoriously argued that pastness, presentness and futurity, though needed to explain the fact of change and thereby give time a direction, cannot be real properties of events. His conclusion was that time itself is not real. It has also been argued that things which do not yet exist, or alternatively, all things which do not exist now, do not, properly speaking, exist at all. These are issues on which a scientific account of time may bear, but they will not be settled simply by noting whether or not our mature scientific theories of time make apparent reference to the objects whose existence is thus questioned.

In his paper, P. C. W. Davies sets the stage for the following three authors by presenting a contemporary physicist's view of time. He begins by distinguishing between those aspects of time about which physics has nothing to say, and those aspects of which a relatively detailed account is available. He includes a brief, provocative statement of the account general relativity gives of the beginning and end of time, before discussing at greater length the asymmetry of physical processes in time. The nature and origins of this asymmetry are, he argues, relatively well understood.

D. H. Mellor presents elements of a philosophical account of time and an analysis of temporal discourse which are in many ways consonant with the physicist's view of time. He denies that events have properties such as being past, present or future, and supports this denial by proposing an account of change, and of the truth-conditions of tensed utterances as well as of their role in our thought. There is a resemblance to McTaggart's argument here. But while McTaggart concluded that time is unreal, Mellor maintains that time itself is real, though certain philosophical views of it are false. In particular, he uses the results of his analysis to criticize an account of the direction of causation given by J. L. Mackie, and a semantic proposal of R. C. Jeffrey, both of which, he argues, involve such views.

The next two papers investigate some relations between the physics and

sions of realism in general. McGinn goes on to propose a non-objectual modal realism, and to examine the problems this raises for our knowledge of modal truths.

By contrast, Bas van Fraassen argues in the final paper of the volume that some central and supposedly objective modal distinctions are, as he puts it, 'but projected reifications of radically context-dependent features of our language'. He himself advocates a view of science which is opposed to that of the scientific realist. In this paper he develops the consequences of his non-realist view for some categories frequently regarded as playing a crucial role in an analysis of science: laws of nature, counterfactual conditionals and essences. It is usually held that science incorporates universal laws of nature, which have been taken to differ from mere accidental generalizations by warranting certain related subjunctive conditional sentences with false antecedents. Van Fraassen maintains that these notions are intimately connected to essences – properties an object must have of necessity. He argues that a formal pragmatics will treat counterfactual conditionals as well as essences as context-dependent, rather as the adverb 'here' varies its reference from one speaker and occasion of utterance to another.

All papers in this volume are here published for the first time. Earlier drafts of the papers were presented to a series of small meetings of philosophers and natural scientists organized by the Thyssen Philosophy Group and financed through generous grants from the Fritz Thyssen Stiftung. On behalf of the Group, I should like to express our thanks to the Fritz Thyssen Stiftung for financial support, and to both the late director, Professor Gerd Brand and the present director, Professor Dr Rudolf Kerscher, for help and encouragement.

Cambridge
February, 1980   Richard Healey

Theoretical explanation

MICHAEL FRIEDMAN

I

A typical, and striking, feature of advanced sciences is the procedure of
theoretical explanation: the derivation of the properties of a relatively
concrete and observable phenomenon by means of an embedding of that
phenomenon into some larger, relatively abstract and unobservable
theoretical structure. Thus, we explain the properties of gases by embedding them into the 'world' of molecular theory: identifying gases with
configurations of molecules, and properties of gases – like temperature –
with properties of molecular configurations – like mean kinetic energy.
We explain the properties of chemical compounds by embedding them
into the 'world' of atomic theory: identifying chemical elements with
particular types of atoms, and properties of chemical elements – like
valence – with properties of atoms – like excesses and defects in the
number of electrons in the outermost shell. And similar examples are
found in all areas of science: the embedding of planetary motions into the
'world' of Newtonian universal gravitation, the embedding of genetics into
the 'world' of molecular biology, the embedding of optical phenomena
into the 'world' of Maxwell's electrodynamics, and so on.

Such theoretical explanations are known in the standard philosophy of
science literature as *reductions*. We read, for example, about the reduction
of the gas laws to kinetic theory, the reduction of chemistry to atomic
physics, and the reduction of optics to electrodynamics. Two kinds of
questions are traditionally discussed in connection with the notion of
reduction. On the one hand, we find relatively technical questions about
the exact logical relationships between the explained or reduced phenomenon and the explaining or reducing theoretical structure. Are the
'identifications' I spoke of above literally *explicit definitions* in the logician's
sense, or is there perhaps some weaker kind of 'identification'? Can we
literally *deduce* the reduced theory from our theory of the relevant higher
level structure, or do we have to rely on auxiliary assumptions, approx-
imations, and other questionable devices? On the other hand, we find
relatively vague and general discussions about whether reduction –

especially when it involves social or psychological phenomena – is a 'good thing'. Here, of course, is where we encounter vigorous debates – usually generating more heat than light – between 'reductionists' and 'anti-reductionists'.

In this paper I would like to focus on a different question – a question that is both obvious and important, but has received surprisingly little attention. Under what circumstances, and to what extent, does the procedure of theoretical explanation or reduction supply us with grounds for believing in the existence or 'reality' of the theoretical structure employed and the truth of our characterization of it? To what extent does a successful theoretical explanation or reduction *confirm* our conception of the theoretical 'world'? Thus, what I want to focus on here is the *point* of theoretical explanation, its function or role in scientific methodology. Specifically, I am interested in how the procedure of theoretical explanation or reduction connects up with (a) the process of confirmation and (b) the postulation of theoretical entities.

I think that my question about the point of theoretical explanation supplies us with some needed perspective for dealing with the two kinds of traditional questions about reduction mentioned above. For first, what the problems that have been raised for the strict, logician's model of reduction (where reduction is seen as a relation of so-called 'relative interpretation' between first-order theories) show is simply that this model is an idealization or first approximation of what actually goes on in science. But if we are to construct a better approximation, we clearly need to know something about the point or function of reduction in scientific method – we need to known what we want our improved model *for*. And second, the broad issue between 'reductionists' and 'anti-reductionists' is surely misconceived. If reduction is construed in a sufficiently abstract sense, i.e. as theoretical explanation, there can be no objection to reduction *per se*. The real issue always concerns *particular* reductions (e.g. the proposed reduction of psychology to neurophysiology) or particular *kinds* of reduction (e.g. micro reduction). Again, however, if we want to know whether particular reductions are good or bad, well-advised or ill-advised, we need some grip on the general function or end of theoretical reduction. At any rate, it is on this assumption that I will proceed.

2

Let us ask, then, about the conditions under which a successful theoretical explanation gives us reason to believe in the theoretical 'world' it employs. Here it is necessary to steer a middle course between two extreme views. The first extreme view, which I will call the *positivist* view, holds that theoretical explanations *never* give us reason to believe in the postulated theoretical 'world'. Theoretical structures should always be seen simply as convenient devices for generating their actual empirical consequences. Indeed, apart from this role, theoretical entities have no 'physical reality' whatsoever.

But why should we take this extreme view seriously? Hasn't positivism been thoroughly 'refuted' in the last thirty years or so? The answer to this is that positivism in its *most general form* has certainly not been 'refuted'. What has been conclusively shown (that is, as conclusively shown as anything ever is in philosophy) is that it is impossible to eliminate theoretical entities by explicitly defining them in observational terms in the manner of Carnap's (1928) *Aufbau*, Russell's (1918) 'logical constructions', or Bridgman's (1927) 'operational definitions'.[1] But elimination via explicit definition in observational terms is not the only way out. It is perfectly possible for the positivistically-minded methodologist to retain theoretical structure – not to try to define it in observational (or other) terms – but, at the same time, to regard such structure as a mere *mathematical representation* of the true empirical facts. On this kind of view, observational or 'phenomenological' facts have the properties they *would* have if they *were* embedded in a larger theoretical 'world', but they are *not* literally so embedded.

To clarify this conception, compare two attitudes that might be taken towards, for example, the molecular model of a gas. If one adopts a realistic attitude towards this model one believes that molecules really exist, that gases really are configurations of molecules, that temperature really is mean kinetic energy, and so on. On the other hand, one can also adopt an anti-realistic attitude which repudiates such 'literalism'. On an anti-realistic conception of kinetic theory there is no real molecular 'world', and gases are not literally embedded in this 'world'. Rather, the function of kinetic theory is to supply a mathematical model for the observable behavior of gases by means of an association or mapping that *correlates* – but does not *identify* – gases and their properties with particular mathematical constructs. Kinetic theory does not literally assert that gases are a particular *sub-part* of a larger 'world' of molecules and their configurations; it simply asserts that there is an association or correlation between the properties of gases and the entities (which can now be viewed as purely mathematical) in such a theoretical 'world'. There is no longer

---

[1] Of course this was well-known to the positivists themselves as early as 1936! Cf. Carnap's 'Testability and Meaning'.

any question of gases really being configurations of molecules or of temperature really being mean kinetic energy.[2]

In general, a realistic conception sees the relationship between theoretical structure and observational phenomenon as one of genuine *reduction:* the observational phenomenon is actually identified with a piece of theoretical structure. An anti-realistic conception sees the relationship as one of *representation:* the observational phenomenon is merely correlated with, or mapped on to, a piece of theoretical structure. As a result, it is no longer necessary to attribute anything more than mathematical 'reality' to theoretical structure. However, since this kind of anti-realistic conception is perfectly coherent, the positivistically minded methodologist can legitimately ask why we should *ever* regard theoretical structure as something more than a representation of the observable phenomena. Why should we ever move from the weaker claim of representation to the stronger claim of an actual reduction? After all, the two claims have precisely the same consequences for observable phenomena alone. What advantage do we gain by making the stronger claim?

This question is made more pressing by the fact that the reduction/representation distinction is not a mere philosopher's invention; it plays a genuine role in scientific practice. Scientists themselves distinguish between aspects of theoretical structure that are intended to be taken literally and aspects that serve a purely representational function. No one believes, for example, that the so-called 'state spaces' of mechanics – phase space in classical mechanics and Hilbert space in quantum mechanics – are part of the furniture of the physical world. Quantum theorists do not believe that the observable world is literally embedded in some huge Hilbert space (unless, perhaps, they adopt the 'many-worlds' interpretation!). Rather, the function of Hilbert space is to *represent* the properties of physical quantities like position and momentum by means of a *correlation* with Hermitean operators. Similar remarks apply to other auxiliary 'spaces' such as 'color space', 'sound space', etc.[3] Positivism

[2] More precisely, the relevant difference between the realistic and anti-realistic conception is between a *submodel* and an *embedding*. If $\alpha = \langle A; R_1, \ldots, R_n \rangle$ is our theoretical structure and $\beta = \langle B; R'_1, \ldots, R'_m \rangle$ $(m \leqslant n)$ is our observational structure, then the realistic conception asserts that β is a submodel of α: i.e. $B \subseteq A$ and $R'_i = R_i/B$ $(i \leqslant m)$. By contrast, the anti-realistic conception asserts only that β is embeddable into α: i.e. there exists a map $h: B \rightarrow A$ such that $h(R'_i) = R_i/h(B)$ $(i \leqslant m)$. Note that this anti-realistic technique for avoiding 'literalism' about theoretical structure is similar to, but more general than, other well-known techniques such as Craig theories and Ramsey sentences. It is more general because it does not assume, as they do, that we have a distinction between observational and theoretical *vocabulary* available. Thus, it avoids other common objections to positivism.

[3] The *relationalist* holds that *physical* space has the same, merely representative, function as these auxiliary 'spaces'. Material objects are not, as the *absolutist* imagines, literally embedded in physical space. The positivist generalizes this relationalist position to *all* theoretical structure.

4

challenges us to find a rationale for this distinction. Why shouldn't all aspects of theoretical structure be treated in the same way? Why not regard the entire theoretical 'world' as one big auxiliary 'space' for *representing* the observable phenomena?

However, in trying to respond to this positivist challenge, we must also be careful to avoid a second extreme view. This view, which has become increasingly popular with the demise of positivism, takes a very liberal and cavalier attitude towards theoretical structure. A theoretical explanation – the embedding of some observational phenomenon in a larger theoretical structure – gives us reason to believe in its postulated structure (or at least provisionally to accept it) whenever it is the *best available* explanation of that phenomenon. So, in particular, we have reason to believe in (or at least provisionally to accept) a piece of theoretical structure that provides the *only* available explanation of some observational phenomenon. This kind of liberal principle, which has the ring of a bland and innocuous truism, plays an important role in contemporary debates about the emerging human 'sciences'. Thus, in both linguistics and philosophy of language the theoretical structure known as possible-world semantics is much in vogue. Its proponents argue that we should take it seriously – and forget the qualms of the skeptics – because, after all, it provides us with the only 'rigorous' explanation of various so-called intensional phenomena (cf. Lewis (1975), Stalnaker (1968), etc.). In contemporary cognitive psychology and philosophy of mind, proponents of a full-blown system of mental representation – a so-called language of thought – argue in strikingly similar ways (cf. Fodor (1975)). Our innocuous-sounding liberal principle is capable of generating some quite dramatic consequences.

The first problem with this extreme liberal principle is that, from a purely abstract point of view, the procedure of theoretical explanation is altogether too *easy*. For, given a theoretical structure of sufficient mathematical complexity (containing a big enough universe of sets), it is possible to provide a model for *any* phenomenon we care to explain.[4] Conversely, given any phenomenon we care to explain, observational or not, it is possible to find a theoretical structure that provides a model for it. Suppose, then, that we start with some observable phenomenon O, and we provide a theoretical model T_1. According to our liberal principle, we accept T_1 (at least provisionally). But now, with sufficient mathematical ingenuity, we can construct a stronger theoretical structure T_2 which models T_1 in the same way that T_1 models O. Should we accept T_2? If so,

[4] This is guaranteed by the completeness theorem, which states that *any* consistent class of sentences has a model in the universe of sets; in fact, in the universe of sets of natural numbers.

5

what about T_3 . . .? Thus, a consistent application of our liberal principle leads to an infinite hierarchy of explanations of ever increasing strength. But this is absurd: virtually *any* theoretical structure will be eventually accepted in this way. More realistic principles of scientific methodology must tell us when it is rational to *ascend* the infinite hierarchy of potential explanations *and* when it is rational to *stop*.

Moreover, the abstract hierarchy of potential explanations has more than one 'dimension'. In trying to explain a particular phenomenon we can look for reductions or theoretical models in more than one direction. Thus, one way to look for a theoretical explanation is to look for a *local* or *micro* reduction. This kind of reduction, of which the kinetic theory of gases is typical, tries to explain the behavior of a class of objects by means of a theory of their composition into smaller *parts*. But this is not the only way to go. It is also often possible to look for a *global* or *macro* reduction which explains the behavior of objects, not in terms of their smaller parts, but in terms of their place in a bigger theoretical *system*. Typical of this kind of reduction is the general relativistic explanation of the law of motion for free falling particles: we identify particles with singularities in the gravitational field and derive the law of motion from Einstein's field equations. However, given some particular phenomenon which we seek to explain, there is no way of knowing *a priori* which direction is best. We need additional methodological principles to guide us.

The trouble with our liberal principle, then, is that it is entirely *too* liberal. The best available explanation may nevertheless be a bad explanation. It may lead us in the wrong direction, and, even if it leads us in the right direction, it may lead us too far. In such circumstances, it may very well be better to have no explanation at all rather than an explanation that leads us astray. Therefore, we need to adopt a more critical attitude towards the best available explanation; and such a critical attitude is all the more necessary at junctures in the history of science – like the present juncture in the development of the human 'sciences' – when the whole problem is knowing what direction to follow. In fact, this kind of juncture provides philosophical criticism with a unique opportunity to do some good. The primary sin of an extreme liberal attitude towards theoretical structure lies in its cavalier dismissal of all such criticism.

3

What we need is an account of theoretical explanation and theoretical structure that is capable of giving us some guidance. We need an account of when it is rational to ascend the abstract hierarchy of potential explanation in a given direction – to accept some particular theoretical structure as a genuine reduction – and when it is rational to go no further in a given direction. We need guidance that is more helpful than the positivist advice to avoid all theoretical structure (except, of course, as mere representation) and the recent liberal advice always to accept the best available structure.

I believe that the key to this account is a property of theoretical structure which has often been noticed in the philosophical literature, although not, I think, with the proper emphasis: namely, its *unifying power*. A good or fruitful theoretical structure does not serve simply to provide a model for the particular phenomenon it was designed to explain; rather, in conjunction with other pieces of theoretical structure, it plays a role in the explanation of many other phenomena as well. Take our old friend the molecular model of a gas, for example. By assuming that gases are composed of tiny molecules subject to the laws of Newtonian mechanics we can explain the Boyle–Charles law for a perfect gas. But this is only a small fraction of our total gain. First, we can also explain numerous other laws governing the behavior of gases: Graham's law of diffusion, Dalton's law of partial pressures, and so on. Second, and even more important, we can integrate the behavior of gases with the behavior of numerous other kinds of objects. Thus, the hypothesis of molecular constitution, in conjunction with atomic theory, for example, helps us to explain chemical bonding, thermal and electrical conduction, nuclear energy, genetic inheritance, and literally hundreds of other phenomena. Similarly, another central set of hypotheses of kinetic theory, Newton's laws of mechanics, figure in the explanation of planetary motion, projectile motion, the tides, etc., etc. In the absence of the theoretical structure supplied by our molecular model, the behavior of gases simply has no connection at all with these other phenomena. Our picture of the world is much less unified.

Now it is extremely important to see that the point of this kind of theoretical unification is not merely aesthetic, but also results in our picture of the world being much better *confirmed* than it otherwise would be. For a theoretical structure that plays a role in theoretical explanations in many diverse areas picks up confirmation from all these areas. The hypotheses that collectively describe the molecular model of a gas of course receive confirmation *via* their explanation of the behavior of gases. But, in addition, they also receive confirmation from all the other areas in which they are applied: from chemical phenomena, thermal and electrical phenomena, nuclear phenomena, planetary phenomena, and so on and so forth. By contrast, the purely 'phenomenological' description of a gas – i.e. the totality of observational laws about gases – receives confirmation from one area only: the behavior of gases themselves. The 'phenomenological' description, unlike the theoretical description, receives no support at all from chemical phenomena, nuclear phenomena, or planetary phenomena. Hence, the theoretical description, in virtue of its greater unifying power, is actually capable of acquiring more confirmation than is the 'phenomenological' description!

Suppose that the process of confirmation works something like this. Each time a theoretical hypothesis figures in the explanation of some observable phenomenon it receives a 'boost' in its degree of confirmation. The size of the 'boost' it receives from any particular explanation in which it figures depends on various factors that are not at all clearly understood; but, at any rate, it is not in general large enough to make our hypothesis rationally acceptable. A fortiori it is not in general large enough to raise the degree of confirmation of our hypothesis above that of a purely 'phenomenological' description of the observational objects whose behavior is explained. However, as time goes on, our hypothesis (if it is a good one) receives additional 'boosts' in degree of confirmation via its use in many other explanations. Eventually (if it is a very good hypothesis) two things happen. First, the degree of confirmation of our hypothesis exceeds that of a purely 'phenomenological' description of the objects in question.[5] Second, our hypothesis receives sufficient confirmation to be rationally acceptable.

Thus, this simplified picture of the process of confirmation accounts for our movement through the abstract hierarchy of potential explanations: we should move to a particular 'place' in this hierarchy – accept a particular piece of theoretical structure – when, and only when, it has received sufficient 'boosts' in its degree of confirmation via its use in many different explanations. This picture also illuminates our earlier distinction between representation and reduction. For representations do not interact with other bits of theory and with each other in the way that reductions do. If we treat a piece of theoretical structure as a mere representation of the observational facts, we take away its unifying power, and, as a consequence, the representation does not receive the repeated 'boosts' in degree of confirmation that we have described.

Suppose we treat the molecular model of a gas as a genuine reduction. We assert that there really is a 'world' of molecules subject to Newton's laws of motion and that gases really are a particular sub-part of this 'world'. As we have seen, this very same theoretical 'world' also figures in many other explanations. For example, in conjunction with the atomic

---

[5] Of course, after the degree of confirmation of our theoretical hypothesis reaches this critical point, the degrees of confirmation of its observational consequences have to be revised upwards as well. The point is that the theoretical description receives more confirmation than the 'phenomenological' description otherwise would have.

theory of molecular structure and the identification of chemical elements with different kinds of atoms, it yields the valence theoretic explanation of chemical bonding. In this way, our initial hypothesis picks up more and more confirmation. On the other hand, suppose we treat the molecular model as a mere representation. We do not assert that there really is a theoretical 'world' of molecules nor that gases really are a sub-part of this 'world'. Instead, we assert only that gases behave *as if* they were embedded in such a 'world'. But this assertion does not figure in further explanations in the way that the reduction does. For example, it no longer figures in the valence theoretic explanation of chemical bonding, because there is no longer any real molecular 'world' that underlies the two explanations. We can assert, if we like, that chemical elements behave as if *they* were embedded in a molecular-atomic 'world', but this assertion has nothing to do with our previous assertion about gases. The common piece of theoretical structure – the molecular 'world' – has been 'trapped' inside the 'as if' clauses.

Diagrammatically, the situation looks like this for a reduction:

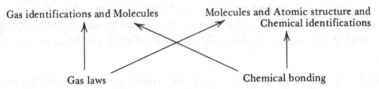

The pieces of theoretical structure in a reduction are free to interact with other such pieces, and to thereby acquire additional 'boosts' of confirmation. On the other hand, the situation looks like this for a representation:

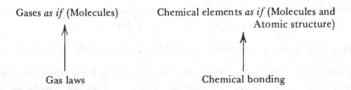

The pieces of theoretical structure in a representation are not 'free', because they are there only to generate 'phenomenological' consequences. As a result, representations receive only the 'boosts' of confirmation appropriate to purely 'phenomenological' descriptions. Representations have no unifying power.

This fact about representations is the answer to our earlier question about when a given theoretical structure should be viewed as a genuine reduction and when as a mere representation. It should be viewed as a genuine reduction – that is, taken literally – when it interacts in a fruitful way with other pieces of theoretical structure: when its unifying power is great. If, on the other hand, it does not interact fruitfully with other pieces of theoretical structure, then there is simply no point in moving from the weaker claim of representation to the stronger claim of reduction. No increase in unifying power – and therefore no increase in confirmation – is thereby effected. In such circumstances, we 'seal off' the theoretical structure within an 'as if' clause. This, if I am right, is what happens with auxiliary 'spaces' such as 'phase space' and 'color space'. Unlike the case of molecules, there is no unifying power to be gained by allowing such auxiliary 'spaces' to populate our theoretical 'world' as autonomous existents. That is why they function only as representations.

But what if some positivistically minded philosopher now suggests that we view the *entire* theoretical 'world' as one *big* representation, as one big auxiliary 'space'? This suggestion is not vulnerable to the problems for representations we raised above, because there is now no question of 'sealing off' some given piece of theoretical structure from fruitful interaction with *other* pieces. Rather, *all* pieces of theoretical structure are to be conjoined together within a *single* 'as if' clause. In other words (assuming for the moment that kinetic theory and atomic theory make up our total theory of the world), we do not formulate our total theory like this:

Molecules and Gas identifications and Atomic structure and Chemical identifications

as the realist would have it. Nor do we formulate our total theory like this:

Gases *as if* (Molecules) and Chemical elements *as if* (Molecules and Atomic structure)

as in our example. Instead, we formulate it like this:

Gases ∪ Chemical elements *as if* (Molecules and Atomic structure).

The point, of course, is that this last formulation is logically stronger than the previous one: ⌜ *as if* (P) and *as if* (Q) ⌝ does *not* imply *as if* (⌜P and Q⌝). In fact, ⌜P and Q⌝ typically has *observational* consequences that are not consequences of P and Q separately.[6]

---

[6] This point about the 'non-conjunctiveness' of representations can be expressed more precisely thus. Let β_1 and β_2 be two structures with common domain B (the observational 'world'). If both β_1 and β_2 are *embeddable* into α, then it does *not* follow that $\beta_1 \cup \beta_2$ is embeddable into α. By contrast, if both β_1 and β_2 are *sub-models* of α, then it does follow that $\beta_1 \cup \beta_2$ is a sub-model as well. Similarly, the Craig theory of ⌜ P and Q ⌝ does not equal the Craig theory of P conjoined with the Craig theory of Q (Craig 1956); the Ramsey sentence of ⌜P and Q⌝ does not equal the Ramsey sentence of P conjoined with the Ramsey sentence of Q (Ramsey 1929). 'Non-conjunctiveness' seems to be a general property of anti-realist strategies. Cf. Putnam (1975) and Boyd (forthcoming).

What is the advantage of the realistic formulation of our *total* theory over the positivistic formulation? If we look at our total theory at a *single* time, no advantage is immediately apparent. All we can say is that the two formulations agree on all observational consequences. However, the situation is radically different if we take account of the evolution of our total theory over time. For the realistic formulation evolves by *conjunction* over time, while the positivistic formulation does not. Suppose at t_1 our total theory is the molecular hypothesis and at t_2 the molecular hypothesis conjoined with a theory of atomic structure. In this case, our theory at t_2 retains an hypothesis from our theory at t_1 which automatically carries with it all 'boosts' in confirmation that it received at t_1. On the other hand, on the positivistic model, our total theory at t_1 is given by

Gases *as if* (Molecules)

while our total theory at t_2 is given by

Gases ∪ Chemical elements *as if* (Molecules and Atomic structure).

Our theory at t_2 does not retain an hypothesis from our theory at t_1; we simply replace one global representation with a stronger one. As a result, the 'boosts' in confirmation that our theory received at t_1 are not automatically carried along at t_2, and our total theory receives fewer 'boosts' in confirmation overall. Therefore, viewed as a strategy for formulating theories across time, the realistic strategy is preferable to the positivistic strategy in all circumstances except one: namely, if some genius invents all the theoretical structure we will ever need at the *beginning* of the process of confirmation. In *these* circumstances the positivistic formulation receives just as many 'boosts' in confirmation as the realistic formulation – but only because it is never either expanded or revised!

4

We have seen enough above, I hope, to give us at least a rough grip on the point or function of theoretical explanation in scientific method. Theoretical structure functions to increase the unity of our overall picture of the world by connecting together many phenomena that are completely unrelated on the observational or 'phenomenological' level. As a result, our overall picture of the world is much better confirmed than it otherwise would be. Moreover, the method of reduction – the procedure of identifying some (relatively observational) objects and properties with other (relatively theoretical) objects and properties – is absolutely essential to this process of theoretical unification. For how else can we unify diverse areas that are described by disjoint vocabularies? Thus, for example, we connect the behavior of gases with the behavior of metals by defining *both* in terms of atomic structure. 'Everything is what it is and not another thing' is not a good slogan for the scientific methodologist.

In general, and in the abstract, therefore, reduction is a 'good thing'. It increases the coherence, the integration, and the confirmation of our scientific theories. For the same reason, I don't see how there can be any objection, in the abstract, to the goal of 'unity of science' – if this means that we should try to embed as many phenomena as possible in a *single* theoretical structure. In practice, however, there is more than one way to go in seeking such theoretical unification. Even given a single theoretical structure – the theoretical structure of contemporary physical theory, for example – there is more than one way to embed any given phenomenon into that structure. The problem for the practising methodologist is to know which way to pursue.

It is precisely here that debates between 'reductionists' – that is, proponents of a particular type or strategy of reduction – and 'anti-reductionists' – that is, opponents of *that* particular type or strategy of reduction – have a point. Moreover, it is here that the conception of theoretical explanation we have developed can perhaps find some useful applications. For we can ask of any particular disputed type or strategy of reduction whether it is likely to lead towards an increase in unifying power, or whether, on the contrary, it is likely to lead away from such an increase. In the former case the given strategy is worth pursuing, and we side with the 'reductionists'. In the latter case the strategy should be abandoned, and we side with the 'anti-reductionists'.[7] I would like to apply these considerations, in what follows, to some debates of this kind in the emerging human 'sciences'. Of necessity, my discussion will be rather sketchy and schematic (even more so than the preceding!).

The first dispute I would like to discuss concerns the status of so-called possible-world semantics. As I mentioned above, this theoretical structure is quite popular in recent linguistic theory and philosophy of language. The basic idea is to regard the actual world as just one member of a larger set of possible worlds, and to capture various intensional notions – like proposition, sense, necessity, belief, and so on – by means of constructions on the set of all possible worlds. Thus, we identify a proposition with the set of all possible worlds in which it is true, the sense of a predicate with a

---

[7] I don't mean to suggest that unifying power is the *only* relevant consideration in deciding between different strategies of explanation. Various considerations – sometimes even ethical and political considerations – are relevant to such a decision. Cf. A. Garfinkel (forthcoming) for an interesting discussion of this matter, especially in regard to social theory.

function from possible worlds to extensions in those worlds, necessity with truth in all possible worlds, etc. Proponents of possible-world semantics feel that these constructions provide an answer to skeptically minded philosophers, like Quine, who have expressed doubts about all such intensional notions. Since they can be constructed within the only available 'rigorous' semantic theory, they are as respectable as any scientific concepts.

Now there is no doubt that possible-world semantics has provided a mathematical *representation* (although an essentially trivial one) for various intensional phenomena. This is just what is shown by the numerous 'completeness' results in 'intensional logic'.[8] But should we also think that it is a genuine theoretical explanation or *reduction,* with ontological claims that should be taken as seriously, say, as the molecular model of a gas? If we think of the postulation of theoretical structure as merely a matter of inference to the best available explanation, we will find this conclusion hard to avoid. We have seen above, however, that this cavalier attitude towards theoretical postulation is quite inadequate. The best available explanation must also lead in the right theoretical direction: it must lead to an increase in unification. How does possible-world semantics fare from this point of view? Not very well, I'm afraid. For, as we have seen, the way a theoretical structure acquires unifying power is through its ability fruitfully to interact with *other* pieces of theoretical structure. But not only are the entities postulated by possible-world semantics completely *sui generis* and unrelated to any theoretical entities that we have previously encountered, they are not even in the same 'world' as the latter. It is extremely difficult, therefore, to see how they *could* relate to the rest of our theoretical structure.

Thus, for example, any serious semantic theory must be eventually connected up with psychological and social theory. It must be possible to show how creatures like us acquire and employ a language with the properties that the semantic theory ascribes to it. It must be possible to show how the theoretical structures assumed in the semantic theory play a role in the actual process of language acquisition and use. If they play no such role, we should be very skeptical of their unifying power. But in possible-world semantics the postulated semantic structure appears to be totally divorced from the psychological process of language acquisition and use. How do *we*, as finite creatures in *this* world, relate to entities (and sets of entities, and sets of sets of entities . . .) in *other* possible worlds? If anything, possible-world semantics makes it harder to connect linguistic

---

[8] It is not very surprising that virtually *any* 'logic' can be represented in possible-world semantics, given its incredibly huge universe of sets!

theory with psychological and social theory than it was before. Possible-world semantics appears to lead away from, rather than towards, theoretical unification.[9]

The second dispute I wish to discuss concerns another theoretical structure that is invoked to explain cognitive and linguistic phenomena. This structure is known in contemporary cognitive psychology and philosophy of mind as a system of mental representation or a language of thought. The idea is to postulate an internal language in which intelligent organisms can formulate beliefs and desires and engage in internal practical reasoning ('computation'). We then explain the behavior of intelligent organisms as the product of such internal practical reasoning. Moreover, we explain the behavior of linguistic organisms – that is, organisms who use an external *public* language – in terms of a *translation* from their external language into their internal one. Here, again, defenders of the controversial theoretical structure in question invoke an inference to the best available explanation: how *else* are cognitive and linguistic phenomena to be explained?

Once again, however, we have to ask about the potential unifying power of our theoretical structure, and, once again, it seems to me that the outlook is not promising. The problem is not, as in the case of possible-world semantics, that the entities postulated are literally 'out of this world'. There is no reason why, for example, the items in the system of mental representation couldn't be neurological states. Rather, the problem is that, at least when the explanation of *linguistic* phenomena is at issue, the postulated theoretical structure is not sufficiently *different* from the observational phenomena to be explained. For we explain linguistic behavior by postulating a *second* language. By contrast, a real explanation reduces the phenomenon to be explained to something completely different; it gives us something new to work with. This is how it is able to connect the phenomenon to be explained with other, apparently quite unrelated, phenomena. Recall the molecular explanation of the behavior of gases and the atomic explanation of chemical bonding. Thus, a real explanation of cognitive and linguistic phenomena – an explanation with genuine unifying power – should relate them to *other* types of phenomena. This end is not served by postulating *more* cognitive and linguistic phenomena inside the head.

---

[9] The only defender of possible-world semantics who is even *aware* of this problem is David Lewis. See his illuminating 'Languages and Language' (1975) for an attempt to deal with the relation between language-as-abstract-semantic-system and language-as-used-by-a-community. Unfortunately, his solution seems to involve the assumption of complicated *pre-linguistic* beliefs and intentions – and so, in practice, it leads directly to the language of thought view discussed below.

If these brief remarks are anywhere near the mark, it follows that there are legitimate grounds for skepticism about many of the theoretical structures that have been proposed in contemporary linguistics, psychology, philosophy of language, and philosophy of mind. There are grounds for doubt that do not depend on a misguided behaviorism, a positivistic distrust of *all* theoretical postulation, or general anti-scientific obscurantism. (Although, to be sure, many of the proponents of this kind of skepticism certainly give the impression that their objections rest on some, or all, of these grounds.) For there are reasons to think that the theoretical structures in question do not fulfil the role or function that good theoretical explanations fulfil. There are reasons to think that the theories in question do not lead us in the right direction in the abstract hierarchy of potential explanations. Either they fail to lead us *towards* an increase in unifying power – as in the case of the language of thought hypothesis; or they lead us definitely *away* from one – as in the case of possible-world semantics.

But can we say anything more positive about these matters? Can we suggest a *better* direction in which to look for explanations in the human 'sciences'? I think that we can. We can try to follow up the emphasis that philosophers like Quine and Wittgenstein have placed on the *social* nature of cognitive and linguistic phenomena. According to these philosophers, human cognitive activities, such as acquiring and applying concepts or affirming and denying judgements, are first and foremost social activities that take place within definite social contexts. If these philosophers are right, we cannot hope to define or explain notions like *concept* or *thought* without explicitly referring to such social contexts. Notions like *concept* or *thought* are not one-place predicates that apply to a single cognitive subject, they are many-place predicates that relate a cognitive subject to a wider community of other cognitive subjects. (This is one way of understanding the notorious 'private language argument'.)

Thus, from this point of view, the problem with contemporary psychology (and philosophy) is not that it tries to give theoretical reductions or explanations of human cognitive activities. There is no objection to theoretical reduction in the abstract. (Although, again, Wittgenstein himself – and many 'Wittgensteinians' – *do* object to theoretical explanation *per se*.) The problem, rather, is the direction of reduction. Contemporary psychology tries to explain *individual* cognitive activity independently from *social* cognitive activity, and then tries to give a *micro* reduction of social cognitive activity – that is, the use of a public language – in terms of a prior theory of individual cognitive activity. The opposing suggestion is that we first look for a theory of social activity, and then try to give a *macro* reduction of individual cognitive activity – the activity of applying concepts, making judgements, and so forth – in terms of our prior social theory. In other words, the correct order or direction of reduction is

social theory

psychology

and not the other way around. From this point of view, the primary problem with contemporary cognitive psychology – along with much else in linguistics, philosophy of language, and philosophy of mind – is a commitment to individualism. Perhaps it is time we turned away from this commitment.

REFERENCES

Boyd, R. (forthcoming). *Realism and Scientific Epistemology*. Cambridge: Cambridge University Press.

Bridgman, P. 1927. *The Logic of Modern Physics*. New York: Macmillan, 1958.

Carnap, R. 1928. *Der Logische Aufbau der Welt*. Berlin. Transl. *The Logical Structure of the World*. Berkeley: University of California Press, 1969.

Carnap, R. 1936.'Testability and Meaning.' *Philosophy of Science* 3, 419–47; 4, 1–40.

Craig, W. 1956. 'Replacement of Auxiliary Expressions.' *Philosophical Review* 65, 38–55.

Fodor, J. 1975. *The Language of Thought*. Scranton, Pa.: Crowell.

Garfinkel, A. (forthcoming). *Explanation and Individuals*. New Haven, Conn.: Yale University Press.

Lewis, D. 1975. 'Languages and Language.' *Minnesota Studies in the Philosophy of Science*, vol. 7, ed. K. Gunderson, pp 1–35. Minneapolis: University of Minnesota Press.

Putnam, H. 1975. 'Explanation and Reference.' *Philosophical Papers*, vol. 2, pp. 196–214. Cambridge: Cambridge University Press.

Quine, W. V. O. 1960. *Word and Object*. Cambridge, Mass.: M.I.T. Press.

Ramsey, F. 1929. 'Theories.' *Foundations*, ed. D. H. Mellor, pp 101–25. London: Routledge, Kegan Paul, 1978.

Russell, B. 1918. 'The Relation of Sense Data to Physics.' *Essays in Mysticism and Logic*. Harmondsworth, Middlesex: Penguin, 1953.

Stalnaker, R. 1968. 'A Theory of Conditionals.' *Studies in Logical Theory*, ed. N. Rescher. Oxford: Blackwell.

Wittgenstein, L. 1953. *Philosophical Investigations*. Oxford: Blackwell.

The case of the lonely corpuscle: reductive explanation and primitive expressions[1]

PETER ALEXANDER

I wish to discuss a problem that I have encountered in working on Locke but which probably has some relevance to the theme of reduction. It concerns the meaning and, more especially, the application of primitive expressions for reductive explanation. In my first section I will say something about reduction in general, in my second section I will outline, without much argument or evidence, the relevant parts of my interpretation of Locke, and in the third section I will discuss my problem and its possible solution. I assume throughout that it is permissible, or even necessary, to talk about properties or qualities.

1. REDUCTION

In this section I make a few points about reduction in order to set the scene for the problem I wish to raise. I assume that there are reductions in the natural sciences that do not necessitate, or even allow, the elimination as 'unreal' or useless of that which is reduced, and it is such reductions that I have in mind. It is not strictly relevant to my theme but I should perhaps say that I believe that Locke's reductions were of this sort.

I take it that one type, if not the only type, of reductive explanation is the explanation of a whole in terms of its parts, ideally, I suppose, its simplest parts, whatever we may mean by that expression. Such explanations are contrasted with what we may call 'holistic' explanations, based upon the view that the parts can be explained or understood only in terms of the whole.[2] One of the historically standard illustrations concerns the

---

[1] I am grateful to David Hirschmann for helping to make this Green Paper less green than it otherwise would have been. He should not be held responsible for the use I have made of his ideas unless he wishes to be. I have also been helped by discussion with members of the Thyssen Philosophy Conference held in September 1978, especially Colin McGinn, Robert Kirk, Michael Friedman, Arnold Zuboff and Hugh Mellor.
[2] I make the common-sense but unanalysed assumption that explanation leads to understanding and that understanding does not always depend upon explanation. Any problems involved here will not, I think, seriously affect my argument. See Friedman 1974, for an interesting discussion.

working of a clock: an explanation that begins with the physical properties of its component bits of metal and makes clear how the hands are turned in terms of their mechanical interactions is contrasted with an explanation that begins with someone's designing the clock to tell the time and makes clear how the physical properties and interactions of the component parts are contrived as they are in order that the whole may serve that purpose. The first implies that the working of the clock can be understood entirely in terms of the properties and movements of its parts, the second that it cannot. However, a more pertinent example, which has the advantage of not involving human designs or purposes and which yet can be used to make the relevant contrast, is physical atomism. Atomism may be regarded as issuing in purely reductive explanations if it is held that its basic concepts can be understood and applied without essential reference to the laws and explanations of the developed theory so that those laws and explanations can be understood entirely in terms of those concepts; it is not regarded as issuing in purely reductive explanations if it is held, in the manner of some philosophers of science, that the 'basic' concepts can be understood only in the light of the developed theory, so that the understanding of both those concepts and the laws and explanations involves, as it were, a constant movement back and forth between them.

This latter view of atomism appears to involve a compromise between purely reductive explanations and fully blown holistic explanations. Even if, in the end, we are forced into some such compromise it is worth exploring the possibility of purely reductive explanation since this appears to have been the ideal of many natural scientists. Robert Boyle, for example, appears to have had this ideal and he was, I believe, followed by Locke in this. Hence my problem.

Ernest Nagel, in his celebrated article 'The meaning of reduction in the natural sciences' (1949) wrote:

The statements of science . . . can be analysed as linguistic structures compounded out of more elementary expressions in accordance with tacit or explicit rules of construction. These elementary expressions . . . may be assumed to have fairly definite meanings fixed by habit or explicit rules of usage. Some of them are familiar expressions of logic, arithmetic, and perhaps higher mathematics but most of them will usually be so-called 'descriptive terms' or combinations of terms which signify allegedly empirical objects, traits, and processes. (1960: 298)

and

Let us refer to those descriptive expressions with the help of which the meanings of all other such expressions may be explicated – whether the explication is given in the form of conventional explicit definitions or through the use of different and more complicated logical techniques – as the 'primitive' expressions of the sciences. (299)

Nagel, surprisingly, says little or nothing about how these primitive expressions are to be given application beyond the mysterious assertion that their meanings are 'fixed by habit or explicit rules of usage'. This is where my problem begins.

It will help to characterize the problem more precisely if I quote from Putnam (1970) the central features of his 'trial-and-error procedure' for introducing 'fundamental magnitude terms', which are, I think, primitive expressions in Nagel's sense. Putnam talks of correlating the properties specific to physics 'with equivalence classes of predicates definable with the aid of fundamental magnitude terms' (1975: 310), and gives the following four conditions for fundamental magnitude terms. They, he says,

(1) 'must be "projectible" in the sense of Goodman';
(2) 'must characterize all things – i.e. all particles in a particle formulation of physics, and all space–time points in a field formulation of physics';
(3) must include '"distance" or a term with the aid of which "distance" is definable';
(4) must allow laws of an 'especially simple form – say differential equations'. (pp. 308–9)

Putnam stresses that we introduce such terms 'not by definition but by a trial-and-error procedure' and this, as is clear from the conditions just quoted, turns out to be a double procedure; the laws are to be discovered at the same time as the fundamental magnitudes. Putnam's conditions particularly interest me because Locke appears to have been guided by similar principles in his theorizing about primary qualities, if my interpretation of his work is correct. I hasten to add, however, that I arrived at my interpretation before I had read this article of Putnam's[3] and that I am not crediting Locke with the anticipation of Goodman on green and grue. Moreover, item (4) in Putnam's list, whatever Putnam intends by it, figures in Locke, and in a very rudimentary form, merely as a means of deciding which qualities should be regarded as primary and not as a means of giving meaning to the primary-quality words; it is unlikely that Locke, or Boyle, thinks that the laws are discovered at the same time as the 'fundamental magnitudes' or that either of them had differential equations in mind. My central problem, however, concerns item (2) of Putnam's list; it is the problem of whether it makes sense to require that our primitive terms characterize all things, that is, whether the fulfilment of the requirement is possible, at least in the way this requirement was

---

[3] This is not to make a claim to priority but to show that my view of Locke was not reached by attempting to apply Putnam's account to Locke.

seen by Boyle and Locke. However Putnam thinks of this requirement, Boyle and Locke thought of it as the requirement that each corpuscle has the primary qualities independently of all the other corpuscles, and it is the question whether this makes sense that I want to explore.

It appears, at least on the face of it, as if this requirement is necessary if we conceive of science as looking for economical theories of as great generality as possible. The best possible atomic theory, for example, would seem to be one in which every physical phenomenon could be explained in terms of just those properties that the atoms must have in order to be material, and on such a theory each atom must have all the primary qualities, whatever they are. Such a theory would have greater systematic power than any theory that allowed some atoms to have properties lacked by others; indeed it would probably have the greatest possible systematic power (Sklar 1967; Hempel and Oppenheim 1948; Kemeny and Oppenheim 1955) and it would embody the ideal of reductive explanation.

However, I think my problem arises whether this is agreed or not, since, even if we reject it, it seems that the very idea of primitive descriptive expressions entails that each of these expressions must apply to some atoms, even if not to all, independently of other atoms or conglomerations of them, if we are aiming at reductive explanations; otherwise, either the explanations cannot get started or they are not purely reductive.

It is perhaps important here to point to a distinction that I am taking for granted. We can talk of the primitive expressions of a particular physical theory, say the theory of electromagnetic radiation, or of the primitive expressions of physical science as a whole. The primitive expressions of a particular scientific theory presumably do not pose the problem that concerns me, since their meanings and applications can be dealt with in terms of other, more fundamental, physical theories; but if we are concerned with what appears to be the ideal of physical scientists, a unifying theory of all physical phenomena, then we *are* faced with the problem.

2. LOCKE AND THE CORPUSCULAR PHILOSOPHY

My unorthodox interpretation of Locke depends largely upon the belief that his work cannot be understood as he intended it unless it is realized that the corpuscular philosophy is absolutely central to it and that a crucial general aim of the *Essay* is to explore the implications of that hypothesis and, indirectly, to support it by showing how much sense can be made of the natural world and our experience of it on this basis. His main influence was, of course, Robert Boyle, in whose works we find much that is usually regarded as specifically Lockean. I have argued in various places for my interpretation (Alexander 1974a, 1974b, 1977)[4] and all I am able to do here is to summarize some relevant parts of it without much support.

Boyle and Locke were dualists and Locke, I believe, was a more radical dualist than Descartes, since he thought that the problem of interaction was in principle insoluble by both scientific and philosophical methods (see e.g. Locke 1690–1706: IV. iii. 13–14, pp. 545–6). He took mental phenomena to be as real as material phenomena. The corpuscular hypothesis was intended to allow mechanical explanations of all material phenomena in terms of the intrinsic properties of the material particles or corpuscles involved in such phenomena. A mechanical explanation was conceived of as one in which all changes are explained by impulse, by the collision of material particles without action at a distance.[5] The properties of these corpuscles are some of those that we associate with matter and, indeed, observe in middle-sized chunks of matter. One active motive was the avoidance of any reliance upon 'occult qualities', that is, qualities that are neither observable nor able to be characterized except in terms of just those effects that they are intended to explain, e.g. the dormitive power of opium. Forces between particles do not figure in this version of corpuscularianism, presumably because they are regarded as occult, so the theory is kinematic rather than dynamic. This is one reason for regarding Boyle as having influenced Locke more than Newton did.

The distinction between primary and secondary qualities is essential to the corpuscular hypothesis, and the distinction between ideas and qualities essential to Locke's use of it and to much of the rest of his philosophy. It is important to take seriously, as I am sure Locke did, the section-heading 'Ideas in the Mind, Qualities in Bodies' (II. viii. 7, p. 134). When we are talking of material things, anything we call a quality is attributable to a material object and anything we call an idea is attributable to a mind; even though we sometimes use the same word for a quality and an idea (e.g. 'extension') we must not suppose that an idea ever *is* a quality of a material object.

Thus both primary and secondary qualities are 'in' objects and neither is mind-dependent. Colours, sounds, tastes and odours are not secondary qualities because they are not qualities; they are ideas (sensations) of, that is, caused by, secondary qualities. Locke frequently says such things as that there are no colours in the dark (e.g. II. viii. 19, p. 139) but it does not

---

[4] Some parts of my interpretation are argued for in work as yet unpublished but clues to most of them will be found in these articles.

[5] Locke takes notice of action at a distance in the *Essay*, 4th and 5th editions, but says that he finds it unintelligible. (Locke, 1690–1706, 135–6, fns.)

follow, and I think Locke never says, that there are no secondary qualities (of the relevant sort) in the dark (Locke, 1690–1706: *passim*). Locke asks, rhetorically, whether the presence or absence of light could make any real alteration in an object (II. viii. 19, p. 139); if colour were a quality, it could. To say that colour is not a *real quality* of an object is not to say that secondary qualities are not real qualities of objects and it is not to say that colour is not real; it is to say that it is not a quality.

However, corresponding to the colour of an object there is a real quality of the object, namely, a specific *texture* responsible for that particular idea. 'Texture' is used as a technical term by Boyle and is adopted in the same sense, I believe, by Locke. The texture of an object is the particular arrangement of corpuscles that constitute it; this texture is causally responsible for its interactions with other things, including light and our bodies, both of which also have textures.

Qualities, whether primary, secondary or 'tertiary' are intrinsic to bodies. (I shall ignore 'tertiary' qualities but it will be obvious how I could bring them in.) The difference between primary and secondary qualities, in this respect, is that primary qualities are intrinsic to every material body, even a single corpuscle, while secondary qualities are intrinsic only to complex bodies comprising collections ('conventions' as Boyle has it) of corpuscles. I want to say, therefore, that secondary qualities *are* textures of such collections; in terms of these textures we can explain, or almost explain – there is still the mind–body problem – the effects of those collections on us.

(This may seem to conflict with Locke's frequent explicit statements that secondary qualities are powers, which appears to imply that they are not intrinsic but relational. I think I can avoid that conflict. Power talk is useful at a certain level but ultimately dispensable; or, more accurately, a geometrical/mechanical, non-dispositional account of powers can be given and plausibly attributed to Boyle and Locke.[6] Any other sense of 'power' would, I think, have savoured of the occult for them. This, however, is not strictly relevant to what follows.)

The importance of the primary/secondary quality distinction is, of course, that the secondary qualities, seen as textures of objects, can, in principle, be described completely in terms of the primary qualities, that is, the qualities of the individual corpuscles and the spatial relations between them. The secondary qualities are analysable into primary qualities and the primary qualities are the primitives for all explanations of physical phenomena. The whole theory is thus designed to issue in purely reductive explanations of the kind I have roughly described.

---

[6] For a sketch of this view see my 'The names of secondary qualities' (1977).

Now the question arises as to how we are to choose our primary qualities. It seemed conceivable to Boyle and Locke that all material phenomena could be explained in terms of material corpuscles. Whether this was indeed so would have to await the carrying out of the programme and the achievement of extensive success in explanation; but embarking on the programme necessitated a decision about which qualities to regard as primary. What could be the criterion for the choice? An appropriate question seemed to be: what qualities are necessary and sufficient for a corpuscle to be *material?* Since a controlling aim was the avoidance of occult qualities, it seemed reasonable to require that the qualities chosen should be *of a kind* with which we are familiar from everyday observation.

If the primary qualities were to be those that anything must have in order to be material, then a single corpuscle, alone in the universe, must have them. Any quality that such a corpuscle need not have need not be said to be primary. This was the central criterion used by Boyle, quite explicitly, and by Locke, less explicitly but, I think, just as surely (Locke, II. viii. 9 and 10, pp. 134–5).

Boyle begins to outline the hypothesis in *The Origin of Forms and Qualities* by referring to 'one catholick or universal matter' that all bodies have in common (1772; vol. III, p. 15), and he says that by matter he means 'a substance extended, divisible and impenetrable'. The impenetrability here referred to is the absolute solidity discussed by Locke and distinguished from empirical hardness (Locke 1690–1706: II. iv). Thus Boyle begins with absolutely solid material corpuscles and asks what qualities they must have. This, I think, is to ask what 'material' involves beyond solidity. He says,

if we should conceive that all the rest of the universe were annihilated, except any [*sc.* any one] of these intire and undivided corpuscles ... it is hard to say what could be attributed to it, besides matter, motion (or rest), bulk, and shape. Whence by the way you may take notice that bulk, though usually taken in a comparative sense, is in our sense an absolute thing, since a body would have it, though there were no other in the world. (p. 22)

and later he says that there are

three essential properties of each intire or undivided, though insensible part of matter; namely, magnitude (by which I mean not quantity in general, but a determined quantity, which we in *English* oftentimes call the size of a body), shape, and either motion or rest (for betwixt them two there is no mean). (p. 34)

These three essential properties Boyle refers to as 'inseparable accidents' of matter or its 'primary affections' and it is clear that they are what Locke calls 'primary qualities'. Thus Boyle quite explicitly holds that the corpuscles are solid and have just three primary qualities which I shall call *shape, size* and *mobility*. 'Mobility' is a word used from time to time by both Boyle and Locke in place of 'motion or rest' and 'motion and rest' when they are referring to the primary quality.

When Boyle, in the first passage quoted, talks about bulk as 'an absolute thing' I do not think that he is using the objectionable notion of absolute size; I think that he means just what he says, namely, that to say that bulk is absolute is to say that 'a body would have it, though there were no other in the world'. A body must occupy some volume even if our specifying the volume depends upon our standards of measurement, which may vary. This is to make a realist point: however we measure, there is something to be measured.

Solidity is in a special position. Boyle, I believe, regards it as essential to matter but not a *quality*, since qualities are what may differentiate portions of matter from one another (1772: vol. III, p. 35). The solidity of the corpuscles is incapable of doing that since it is the absolute solidity of whole matter, so all corpuscles are equally solid. Observable hardness is different; it 'admits of degrees' because it applies only to bodies composed of corpuscles plus empty spaces. These are two different, even if distantly related, concepts. I believe that Locke follows Boyle, and perhaps Aristotle, in his conception of qualities, although he does not say so as clearly as one would wish (Locke 1690–1706: II. xxiii. 8, p. 300 and II. viii. 26, pp. 142–3).

So, for Boyle, the idea of a solitary corpuscle in otherwise empty space is of great importance; he asks what qualities we cannot conceive of such a corpuscle as lacking and, by implication, what qualities we need not conceive of it as having. I believe that a close examination of Locke's definition of primary and secondary qualities (Locke 1690–1706: II. viii. 9–10, pp. 134–5) makes it plausible to attribute the same approach to him. It is of the utmost importance to note (*pace* almost everyone) that Locke does not use the idea of resemblance in *making* the distinction; that idea does not occur, relevantly to this, until section 15, when he has moved on to talk of something else. The distinction concerns the way the world is; resemblance enters in the consideration of our knowledge of it.

Now it seems that there are at least two conditions to be satisfied in the choice of primary qualities.

(1) They must be adequate for the ultimate explanation of every physical phenomenon.
(2) They must be intelligible prior to any explanation based on them; that is, their meaning must not be dependent upon the completed system of explanations.

The second requirement appears to be involved in the conception of the attribution of the primary qualities to the lonely corpuscle. I shall return to this. The first requirement involves the possibility of looking forward and, without arriving at detailed explanations, forming some idea of what is needed for the explanation of various phenomena in the proposed manner. This is perhaps a primitive form of Putnam's item (4) quoted above. Roughly, we can say that no quality need be attributed to the corpuscles if it seems plausible to suppose that it might be analysed into or explained in terms of other qualities and if its absence would not detract from the materiality of the corpuscles, as the absence of shape, size or mobility would. This procedure is not part of the process of giving meaning to the quality-expressions accepted and it does not imply that meaning can be given to them only in the light of the laws and explanations that they make possible; it is merely part of the process of deciding which qualities must be attributed to individual corpuscles and which need not. It does not, therefore, militate against the possibility of purely reductive explanations. It might be said, indeed it has been said, that being solid *entails* being coloured just as it entails having size, so there is no possibility of excluding colour as a primary quality. I think a distinction can be made. If being solid entails having size it does so independently of any epistemological considerations, but I think that to show that being solid entails being coloured epistemological considerations must be used. Whether epistemological considerations *can* be dispensed with is not the concern of this paper.

Absolute space, conceived of as providing a frame of reference, was unintelligible to Boyle (1772: vol. III, p. 22) and Locke (1690–1706: II. xiii, pp. 166–81), so our solitary corpuscle could not have a position; but that was acceptable because position could be derived from size, involving distance, as soon as more corpuscles were present, and it only then becomes applicable.

We can sensibly attribute shape, size and mobility to unobservable particles because we are merely attributing qualities of kinds that we observe in middle-sized bodies; the qualities are no different but the bodies are much smaller. Solidity presents something of a problem in this respect since absolute solidity is never observable, but that need not concern us here.

The situation, then, seems to be this. The expressions for primary qualities are understood by us, acquire their meaning, through ordinary, everyday, pre-theoretical observation. We have selected from observable qualities a small number which we think will serve to account for all the rest and which we regard as necessary and sufficient for something's being material. We are hoping for the most general theory of physical phenomena based upon corpuscles having just these primary qualities, shape, size and mobility. Now an essential prerequisite for the construction of our theory, or even for the provision of one explanation of the required sort, is that it makes sense to attribute these primary qualities even to one corpuscle alone in the universe. My question is '*Can* we make sense of this?' The criticisms I have most frequently met in putting forward my view of Locke are based upon the belief that we cannot.

I think that this is not just a little local difficulty since I do not see how any theory that aims at purely reductive explanation can avoid at least raising the question. What poses the problem is not the particular expressions chosen as primitive but the idea of primitive expressions itself. I am inclined to think that the more complex the primitives of our theory are, the more difficult it is likely to be to answer the question, but that at the moment can be no more than a hunch. My impression, based upon a great deal of ignorance and probably as much misunderstanding, is that the most recent views about fundamental particles in physics would afford a fruitful area in which to press the question.

3. SHAPE, SIZE AND MOBILITY

In this section I consider what hope there is for answering my question as it arises in connection with the corpuscular hypothesis as it was understood by Boyle and Locke. That is, I ask whether it makes sense to attribute shape, size and mobility to a single corpuscle alone in the universe. If these are taken to be, or even to be *among*, the defining properties of matter, then the question becomes, I suppose, 'Does it make sense to say that a single corpuscle, alone in the universe, would be material?' Intuitively it seems to make sense if we are prepared to regard matter as a fundamental category at all. Whatever analysis we give of material objects, it seems likely to involve fundamental entities of some sort, whether corpuscles or more mysterious sorts of particles or space-time points. It would seem to be contingent that the universe contains the exact number of these that it does and so contingent that it contains more than one. If it contained only one that one would have to have some properties. (One what?) Or, to return to Lockean terms, it seems reasonable to say that if one corpuscle which among its fellows had shape and size were, by some cosmic catastrophe or act of God, left alone, it would still have shape and size even if no specification of its shape and size were possible.

It might be thought that since we are considering a universe of many corpuscles between which collisions occur, and so on, their being solid entails their having shape, size and mobility, so if we can show that it makes sense to say that our lonely corpuscle is solid we have done all that is necessary. However, that entailment holds, if it holds at all, I think, only if the stated conditions are satisfied. Someone might claim that the universe is just one infinite solid. Showing that our corpuscle has size, shape and mobility contributes to showing that it makes sense to talk of a universe of many finite corpuscles. If we could show that our corpuscle could intelligibly be said to be solid but not to have shape that would presumably be to show that it could not intelligibly be said to be finite. That, in turn, would perhaps be to show that we could not make sense of our universe in terms of material atoms. So it seems necessary to consider the qualities separately.

I begin with size, which seems difficult because of its essentially quantitative and comparative character. For a thing to have size at all it seems essential that there be something else which it is larger than, smaller than or equal to, or by reference to which its size can be precisely specified. Without this we seem to be using the idea of absolute size, which is wicked.

Perhaps a distinction is possible here, as I hinted in talking about Boyle. The objection to absolute size is that it suggests that a body can have a specific size which is in principle unspecifiable, since any specification would be by means of some arbitrary measuring device and, *ex hypothesi*, no arbitrary measuring device can get at the absolute size. However, it may be that to say that our lone corpuscle has size need not involve this. Could we not hold that this is to say that it is the sort of thing for which a size would necessarily be specifiable if other corpuscles were present, that is if we had some standard of measurement? We are talking of a property it now has by virtue of which a certain sort of measurement, under the appropriate conditions, would necessarily give a result. This is not a dispositional property in any ordinary sense, because to attribute it is not to say that under certain conditions the corpuscle would behave in a particular way or change in a particular way but merely that a property it now has would appear, or be 'revealed', in some way or other depending upon how we looked at it.

We might, alternatively, approach this through solidity. Even though a body in otherwise empty space cannot be said to have a position, it can surely be said to occupy a portion of space. Think of space as full of corpuscles. Now think of corpuscles being annihilated until just two are left; there is no difficulty about attributing size to them at least in a rudimentary sense. Now suppose that one of them is annihilated; can we not say that the remaining one occupies a portion of space as it did before? If the body is absolutely solid, some portion of space is not available for further occupancy; if another corpuscle were introduced there would necessarily be two or, since the corpuscles might come into contact, there would be two exemplifications of shape, e.g. two spheres touching. Perhaps we can say that if something has two non-coincident points on it then it has size; if it has some shape then that size is finite.

27

I now turn to shape. I think it may be possible to make sense of our lone corpuscle's having a particular shape through geometry alone without reference to anything external to the corpuscle. The idea of distance is available since to be solid is to have some volume and we can conceive of points at different distances from one another within any volume. We can also give sense to direction purely internally and geometrically. Then for a corpuscle to have a spherical shape is for there to be a point, x, within it such that in every direction there is a farthest point on the solid such that all these farthest points are equidistant from x. If this condition is not fulfilled then the corpuscle may have some other shape which could be defined along similar lines, although the definition would be more complex. Even this allows us to say that we could give a description of a single-corpuscle universe, if that corpuscle is spherical, which would differ from its description if the corpuscle were any other shape. This no doubt relates to Putnam's view that distance must be among the fundamental magnitudes, since it is an attempt to deal with shape in terms of distance and direction.

I now consider mobility. Boyle says something that may be construed as a reason for his sometimes using 'mobility' instead of 'motion or rest'. It is not essential to matter either that it be in motion or that it be at rest. Matter, he says, 'is as much matter when it rests as when it moves' (1772: vol. III, p. 15). However, in any actual situation in the world as it is, a body must be in some state of motion (including the zero state). When Boyle and Locke use 'mobility' in this context they must be using it as a technical term, since its normal use is to say of something at rest that it could move or be easily moved, but we cannot say this of our solitary corpuscle. There appears to be an insuperable difficulty; if to say that it has mobility is to say that it is in some state of motion then it seems impossible to say that it has mobility, because there is nothing by reference to which it could be in some state of motion; if to say that it has mobility is to say that it *could* be in some state of motion, then it seems impossible to regard mobility as an intrinsic property because in its present isolated state it could not be in some state of motion, so its having mobility would seem to depend upon its relations to some other corpuscle.[7] This has seemed an insuperable difficulty at least to most of my critics. I hope to overcome it.

There are signs that the problem worried contemporaries of Boyle. One argument that was used was this. Suppose that there are just two corpuscles in empty space and that they are moving apart. If there is no

[7] This would no doubt be an appropriate place at which to consider Hugh Mellor's 'In defence of dispositions' (1974), but that would involve too great a change of direction for the present paper.

causal interaction between them then the annihilation of one could make no difference to the other; so if the remaining one was at rest before the annihilation of the other it must still be at rest after the event, and similarly if it was in motion.[8] This clearly will not do – but need I go on?

It might seem that rotary motion was more promising. Could we say that our lone corpuscle was rotating about an axis? For simplicity consider a spherical corpuscle. If there is a point on its surface that is tracing a path relative to some other point, e.g. its geometrical centre, then the sphere is rotating – if not, not. This will not do because the idea of tracing a path depends upon the idea of a change of direction of a line joining two points and that requires an external point of reference; in this case no internal specification of direction will do, if we are dealing with a rigid body, because it will not give a *change* of direction.

It will not do either to attempt to deal with this in a Newtonian way with the help of forces, centrifugal or centripetal. Apart from the fact that this move is not open to Boyle and Locke, it is not clear how this could solve the problem for a rigid body without the assistance of at least one other body to be attracted or repelled by the corpuscle in question.

All this appears to reinforce the view that to make sense of a thing's having mobility we need an external point of reference. However, this may be because it is difficult to avoid thinking of mobility as the ability or capacity of a body to behave in a certain way under certain conditions, that is, in the presence of another body, and that may not be the only way of thinking of it. In trying to find intrinsic properties of matter we are looking for properties that define the kind of stuff it is in the most general sense, and that means properties that allow kinds of behaviour, or kinds of properties, each kind being such that specific exemplars of it are inconsistent with one another and such that any particular portion of matter must exhibit some specific exemplar in any circumstances with which we have to deal. In other words, each defining property allows a disjunction of specific properties no two disjuncts of which can be consistently attributed to any one body at the same time. (I am avoiding the terms 'determinable' and 'determinate' for a reason that will emerge.)

Mobility, in the present sense, is a property partly defining a kind of stuff as being, necessarily, at rest or in motion in any situation we encounter. To say that a single corpuscle has mobility is to say that it has the property that ensures that in any actual situation it will be in some state of motion. It would not make sense to say of something that didn't have mobility, say an idea, that it was at rest or that it was in motion. Having mobility is not the same property as either of these; being at rest

---

[8] An early example of the use of this argument is to be found in the works of Marsilius of Inghen in the fourteenth century. See Clagett 1959: 16.

implies having mobility but having mobility does not imply being at rest, and similarly for being in motion. A body's being in motion is its changing position with respect to something, its being at rest is its not changing position with respect to something; its having mobility is not either or both of these. Its having mobility is its being such that it must either be changing or not changing position if there are positions to change. It is being such that if there are positions it must be in one of them.

We should not think of mobility as if it were comparable to solubility and other dispositional properties. To attribute solubility to something is to say that, certain conditions being fulfilled, it will dissolve; it is not to say that in any actual situation it will either dissolve or not dissolve. To attribute mobility to something is not to say that, certain conditions being fulfilled, it will move; it is to say that in any actual situation it will be either at rest or in motion. To attribute mobility to a lone corpuscle is to say that given any conditions other than that of its being alone it will have certain dispositional properties which will be exhibited under certain further conditions e.g. that of moving when pushed or stopping when resisted. Mobility is more like immersibility in a liquid than it is like solubility. To attribute mobility or immersibility in a liquid to something is part of saying that it is material. To say of something that it is material is to say that it is the sort of thing of which it makes sense to say that it must move or not move, must dissolve or not dissolve, must vapourize or not vapourize, must oxidize or not oxidize, and so on, through a whole range of dispositions. Solubility is a property of some kinds of matter; mobility is a property of all matter whatsoever.

This is not to say something empty. Corpuscles, whether lonely or gathered together in conventions, are neither cheerful nor not cheerful, emotional nor unemotional, or ambitious nor unambitious. (It is worth considering why there is something called the materialist theory of mind.) Locke, being a convinced dualist, devotes a good deal of space in his *Correspondence with Stillingfleet* (1697–9) to distinguishing between material substance, which is solid, and mental substance, which is unsolid; material things are such things as have shape, size and mobility, mental things are such things as don't have those qualities but do have others. We are here dealing with the most fundamental categories for characterizing stuffs.

The view I have been putting forward appears to pose difficulties for the idea that mobility is a *quality*, given the conception of quality that I outlined. Is mobility capable of differentiating one portion of matter, one corpuscle, from another? The contrast with which I was concerned when I raised this was with solidity, absolute solidity. There are no degrees of solidity; degrees of empirical hardness are not varying appearances or manifestations of solidity. These are different concepts, solidity involving the absence of empty space from a body, hardness involving its presence. In any situation in which the differentiation of bodies is possible, mobility is being manifested in particular states of motion, just as shape and size are being manifested in particular shapes and sizes. That is, the three primary qualities admit, even if a little indirectly, of degrees in any actual situation, and this will serve for differentiation.

It might seem appropriate, if I am right, to regard shape, size and mobility, applied to matter as such, as determinables or genera, and particular shapes, sizes and states of motion, manifested in particular bodies, as determinates or species. However, this does not seem right since red, a paradigm of a determinate, is a kind of colour, a paradigm of a determinable; and horse, a paradigm of a species, is a kind of *Equus*, a paradigm of a genus, but a particular state of motion is not a kind of mobility.

Perhaps it would be more appropriate to talk of first-order and second-order properties. I have argued that mobility is a different property from either being in motion or being at rest; a corpuscle's having mobility is its having a property such that under certain conditions it must have the property of being in a specific state of motion. Putnam gives as an example of a second-order property that of having a certain machine table about which he says that it is '*a property of having properties which* . . .' and this, 'although a property of the first level (a property of things), is of "second-order" in the old Russell–Whitehead sense, in that its definition involves a quantification over (first-order) physical properties' (1975, p. 313). There appears to be a formal similarity here with what I have said about mobility and what I may have to say about size.

(Locke shows signs, in the *Correspondence with Stillingfleet,* of allowing as qualities anything that differentiates material things from mental things; this would seem to be the idea of second-order qualities.)

However, this exposes another problem. In arguing that it makes sense to attribute mobility to a single corpuscle alone in the universe I have been forced into saying that although it does make sense to attribute mobility it does not make sense to attribute to it the very same property, a particular state of motion, that we attribute to an actual corpuscle and that figures in reductive explanations. How serious a problem is this?

In the first place, one aspect of my problem was the problem whether *any properties at all* could be attributed to the lone corpuscle. If I am right then the answer is that some properties can. I have been covertly asking whether it makes sense to start with material corpuscles at all. In order for this to make sense it seems to me that some properties must be attributable to matter as such, and so to each corpuscle, to distinguish matter from anything else that the universe contains, or might be said to contain. Shape, size and mobility, as explained, appear to suffice for this.

My understanding of Putnam's item (2) is that it is meant to ensure that the primitives of any reductive explanation should be attributable to the basic objects of any such explanation. The kind of explanation I am considering is in terms of material corpuscles; and explanation is called for, and possible, only when there are collections of these, so any explanation must start from some corpuscles among many. In that situation we have, and start from, particular shapes, sizes and states of motion. However, any second-order properties that each of these corpuscles has are such as to ensure the possession by the relevant corpuscles of just those primary qualities, and no others, that figure in any actual explanation.

Having arrived at this point I am struck by the untidiness of the view which I find myself entertaining. The lone corpuscle can be said to have one first-order property, shape, and one second-order property, mobility, and one property, size, which I have left hovering uneasily between the two. However, the whole point of considering the lone corpuscle was to consider what can be said about matter as such. If second-order properties are acceptable at all then presumably there is no objection to saying about our lone corpuscle, or matter as such, that it has the second-order properties shape, size and mobility. To show that it makes sense to say that it has some particular shape (first-order) is not to show that it does *not* make sense to say it has the second-order quality, shape. Nothing needs to have, in order to be material, any particular shape; it needs to have some shape or other. So perhaps I can say that the lone corpuscle has the second-order properties shape, size and mobility, that is, properties such that in any actual situation corpuscles-in-convention would each have the first-order properties, a particular shape, a particular size and a particular state of motion.

The situation, therefore, is this. We understand particular shapes, sizes and states of motion through everyday observation. We apply these without difficulty to the unobservable corpuscles postulated as occurring in collections in actual phenomena. We understand what it is to say that these corpuscles are material by attributing to each the second-order properties shape, size and mobility and these we understand in terms of the corresponding first-order properties. All our explanations start from these first-order properties so these are the properties that are primary for explanation. The second-order properties are primary in the sense that they characterize, in the most general way possible, the kind of stuff we are dealing with, and that they indicate what all the phenomena have in common in spite of wide differences between the observed properties of those phenomena.

It is worth, finally, elaborating a little some of the points I have made in order to avoid objections that have been made in discussion and others that I expect to be made.

It is important to distinguish between Euclidian geometrical theory and the physical theory with which Boyle and Locke were mainly concerned. Euclidian geometry was available in advance of the postulations on which the physical theory was based. I have been suggesting that shape, for example, is an undefined primitive of the physical theory, and my argument has been an informal one, based on the available geometry, for the attribution of a shape to each corpuscle independently of the others. This is not to replace corpuscles by points as the fundamental entities of the physical theory; the fundamental entities remain corpuscles without physically separable parts. Points are not separable parts of corpuscles as corpuscles are separable parts of physical objects. Corpuscles play an explanatory role in relation to physical phenomena. But points, if only because they are merely geometrical segments rather than physically separable parts of corpuscles, cannot be causally related to the corpuscles of which they are segments. I suggest then that my argument does not commit me to an ontology of points rather than corpuscles or even to the addition of points to the ontology of corpuscles. (Perhaps I can say that points are part of the ontology of geometry, not of physics.)

It has been argued that primitive predicates need not be monadic; relational predicates are acceptable and may be essential. The theory under examination is that the predicates applicable to corpuscles are monadic. This is not to say that all primitive predicates used in the physical theory need be monadic, and I have not intended to say this, but that some must be. In so far as the *applicability* of monadic predicates to single corpuscles needs to be justified in geometrical terms, relational predicates such as distance may have to be used, but not in the form of distances between corpuscles.

If it is argued that *all* primitive predicates might be relational, with the consequence that nothing could be said about a universe containing only one corpuscle, I can only suppose that that might be possible but I must add that I am not talking about such a theory. If we are considering a universe of contingently related corpuscles then it is contingent that it contains more than one corpuscle, so it makes sense to consider a universe containing only one. It follows that it makes sense to say that some predicates apply to it and they must be monadic.

I have suggested that immersibility in a liquid is a second-order property. I introduced the idea in connection with solubility and it may, in consequence, be thought that I am suggesting that immersibility must be defined in terms of solubility and insolubility. That is not my intention. Solubility and insolubility can, of course, characterize only something that is immersible, but the first-order properties corresponding to immersibility are 'being immersed' or 'being not immersed' in a liquid. If we have a universe containing just one corpuscle and then introduce a collection of corpuscles constituting a liquid then the original corpuscle *will be* either immersed in it or not immersed in it. This may not be exactly analogous to mobility but it certainly makes mobility look more like immersibility than like solubility, which was my aim.

Do my arguments have the uncomfortable implication that the world is full of second-order properties corresponding to every disjunction of specific properties and, for the most part, doing no work? For example, would my arguments not allow me to say that the lonely corpuscle had position as a second-order property? Perhaps they would. Perhaps, indeed, Boyle and Locke were wrong to suppose that it was unnecessary to attribute position, in that sense, to single corpuscles. However, such attributions could be restricted to properties that could be attributed to the corpuscle itself in the presence of other corpuscles. Thus *it* would be in motion or have a position under certain conditions but *it* would not have a texture, no matter what collection of corpuscles it was a member of, since texture is attributable only to collections; the *collection* would have a texture but not each member of it. Thus there are many properties that could be attributed to conventions of corpuscles or corpuscles-in-convention without attributing corresponding second-order properties to the constituent corpuscles independently; for example, being attracted or repelled by, being soluble in and reacting chemically with. The only properties which would require the attribution of second-order properties would be the defining properties of matter. I have been concerned not with the questions whether Boyle and Locke attributed enough properties or too many or the right ones but with the question whether the attribution of any properties to a lonely corpuscle makes sense. That is, I have been considering the intelligibility of the project rather than the adequacy of their particular pursuit of it.

4. CONCLUDING UNPHILOSOPHICAL POSTSCRIPT

I am not overwhelmed with confidence that I have solved my problem or, indeed, that it is a problem of any importance outside the primitive kind of atomic theory I have been considering. My approach has been largely determined by the historical considerations that led me to the problem and my efforts to rehabilitate Locke. Boyle's and Locke's primitive expressions form perhaps the most rudimentary basis for physical theory and, as history has shown, explanatory adequacy appears to require the addition of further primitive expressions. On the face of it, however, as long as we talk about fundamental particles it looks as if the problem will present itself in some form. On the other hand it may be that physics has departed so far from the basic features of classical atomism that it no longer makes sense to regard fundamental particles as capable of having any properties in isolation from one another. I suspect a tension between the continued talk of fundamental particles and other aspects of recent physical theory, but that is a subject for another paper and another author. It may be that I have been engaged in a *reductio* of purely reductive explanation, or of matter as a starting point, or both.

REFERENCES

Alexander, P. 1974a. 'Curley on Locke, and Boyle.' *Philosophical Review* 83, 229–37.

Alexander, P. 1974b. 'Boyle and Locke on primary and secondary qualities.' *Ratio* 16, 51–67.

Alexander, P. 1977. 'The names of secondary qualities.' *Proceedings of the Aristotelian Society* 77, 202–20.

Boyle, R. 1772. *Works* (ed. Birch), vol. III: *The Origins of Forms and Qualities*.

Clagett, M. 1959. *The Science of Mechanics in the Middle Ages*. Madison: University of Wisconsin Press.

Friedman, M. 1974. 'Explanation and scientific understanding.' *Journal of Philosophy* 71, 5–19.

Hempel, C. and Oppenheim, P. 1948. 'The logic of explanation.' *Philosophy of Science* 15, 135–75.

Kemeny, J. and Oppenheim, P. 1955. 'Systematic power.' *Philosophy of Science* 22, 27–33.

Locke, J. 1690–1706. *An Essay Concerning Human Understanding*, ed. P. Nidditch. Oxford: Clarendon Press, 1975.

Locke, J. 1697–9. *Correspondence with Stillingfleet*. Collected Works, vol. iv. London, 1801.

Mellor, D. H. 1974. 'In defence of dispositions.' *Philosophical Review* 83, 157–81.

Nagel, E. 1949. 'The meaning of reduction in the natural sciences.' *Science and Civilization*, ed. Robert C. Stauffer. Madison: University of Wisconsin Press. Quoted here from *Philosophy of Science*, ed. A. Danto and S. Morgenbesser. Meridian Books, 1960.

Putnam, H. 1970. 'On properties.' *Essays in Honor of Carl G. Hempel*, ed. N. Rescher. Dordrecht: Reidel. Quoted here from Putnam, *Philosophical Papers*, vol. I. Cambridge: Cambridge University Press, 1975.

Sklar, L. 1967. 'Types of inter-theoretic reduction.' *British Journal for the Philosophy of Science* 18, 109–24.

Three kinds of intentional psychology[1]

D. C. DENNETT

I

Suppose you and I both believe that cats eat fish. Exactly what feature must we share for this to be true of us? More generally, recalling Socrates' favourite style of question, what must be in common between things truly ascribed an *intentional* predicate – such as 'wants to visit China' or 'expects noodles for supper'?[2] As Socrates points out, in the *Meno* and elsewhere, such questions are ambiguous or vague in their intent. One can be asking on the one hand for something rather like a definition, or on the other hand for something rather like a theory. (Socrates of course preferred the former sort of answer.) What do all magnets have in common? First answer: they all attract iron. Second answer: they all have such-and-such a microphysical property (a property that explains their capacity to attract iron). In one sense people knew what magnets were – they were things that attracted iron – long before science told them what magnets were. A child learns what the word 'magnet' means not, typically, by learning an explicit definition, but by learning the 'folk physics' of magnets, in which the ordinary term 'magnet' is embedded or implicitly defined as a theoretical term.[3]

Sometimes terms are embedded in more powerful theories, and sometimes they are embedded by explicit definition. What do all chemical elements with the same valence have in common? First answer: they are disposed to combine with other elements in the same integral ratios.

[1] I am grateful to the Thyssen Philosophy Group, the Bristol Fulbright Workshop, Elliot Sober and Bo Dahlbom for extensive comments and suggestions on an earlier draft of this paper.

[2] Other 'mental' predicates, especially those invoking episodic and allegedly *qualia*-laden entities – pains, sensations, images – raise complications of their own which I will not consider here, for I have dealt with them at length elsewhere, especially in *Brainstorms* (1978). I will concentrate here on the foundational concepts of belief and desire, and will often speak just of belief, implying, except where I note it, that parallel considerations apply to desire.

[3] The child need learn only a portion of this folk physics, as Putnam argues in his discussion of the 'division of linguistic labour' (1975).

Second answer: they all have such-and-such a microphysical property (a property which explains their capacity so to combine). The theory of valences in chemistry was well in hand before its microphysical explanation was known. In one sense chemists knew what valences were before physicists told them.

So what appears in Plato to be a contrast between giving a definition and giving a theory can be viewed as just a special case of the contrast between giving one theoretical answer and giving another, more 'reductive' theoretical answer. Fodor (1975) draws the same contrast between 'conceptual' and 'causal' answers to such questions, and argues that Ryle (1949) champions conceptual answers at the expense of causal answers, wrongly supposing them to be in conflict. There is justice in Fodor's charge against Ryle, for there are certainly many passages in which Ryle seems to propose his conceptual answers as a bulwark against the possibility of *any* causal, scientific, psychological answers, but there is a better view of Ryle's (or perhaps at best a view he ought to have held) that deserves rehabilitation. Ryle's 'logical behaviourism' is composed of his steadfastly conceptual answers to the Socratic questions about matters mental. If Ryle thought these answers ruled out psychology, ruled out causal (or reductive) answers to the Socratic questions, he was wrong, but if he thought only that the conceptual answers to the questions were not to be given by a microreductive psychology, he was on firmer ground. It is one thing to give a causal explanation of some phenomenon and quite another to cite the cause of a phenomenon in the analysis of the concept of it.

Some concepts have what might be called an essential causal element.[4] For instance, the concept of a genuine Winston Chruchill *autograph* has it that how the trail of ink was in fact caused is essential to its status as an autograph. Photocopies, forgeries, inadvertently indistinguishable signatures – but perhaps not carbon copies – are ruled out. These considerations are part of the *conceptual* answer to the Socratic question about autographs.

Now some, including Fodor, have held that such concepts as the concept of intelligent action also have an essential causal element; behaviour that appeared to be intelligent might be shown not to be by being shown to have the wrong sort of cause. Against such positions Ryle can argue that even if it is true that every instance of intelligent behaviour is caused (and hence has a causal explanation), exactly *how* it is caused is inessential to its being intelligent – something that could be true even if all intelligent behaviour exhibited in fact some common pattern of causation.

[4] Cf. Fodor 1975: 7n.

That is, Ryle can plausibly claim that no account in causal terms could capture the class of intelligent actions except *per accidens*. In aid of such a position – for which there is much to be said in spite of the current infatuation with causal theories – Ryle can make claims of the sort Fodor disparages ('it's not the mental activity that makes the clowning clever because what makes the clowning clever is such facts as that it took place out where the children can see it') without committing the error of supposing causal and conceptual answers are incompatible.[5]

Ryle's logical behaviourism was in fact tainted by a groundless anti-scientific bias, but it need not have been. Note that the introduction of the concept of valence in chemistry was a bit of *logical chemical behaviourism*: to have valence *n* was 'by definition' to be disposed to behave in such-and-such ways under such-and-such conditions, *however* that disposition to behave might someday be explained by physics. In this particular instance the relation between the chemical theory and the physical theory is now well charted and understood – even if in the throes of ideology people sometimes misdescribe it – and the explanation of those dispositional combinatorial properties by physics is a prime example of the sort of success in science that inspires reductionist doctrines. Chemistry has been shown to reduce, in some sense, to physics, and this is clearly a Good Thing, the sort of thing we should try for more of.

Such progress invites the prospect of a parallel development in psychology. First we will answer the question 'What do all believers-that-*p* have in common?' the first way, the 'conceptual' way, and then see if we can go on to 'reduce' the theory that emerges in our first answer to something else – neurophysiology most likely. Many theorists seem to take it for granted that *some* such reduction is both possible and desirable, and perhaps even inevitable, even while recent critics of reductionism, such as Putnam and Fodor, have warned us of the excesses of 'classical' reductionist creeds. No one today hopes to conduct the psychology of the future in the vocabulary of the neurophysiologist, let alone that of the physicist, and principled ways of relaxing the classical 'rules' of reduction have been proposed. The issue, then, is *what kind* of theoretical bonds can we expect – or ought we to hope – to find uniting psychological claims about beliefs, desires, and so forth with the claims of neurophysiologists, biologists and other physical scientists?

Since the terms 'belief' and 'desire' and their kin are parts of ordinary language, like 'magnet', rather than technical terms like 'valence', we must first look to 'folk psychology' to see what kind of things we are being asked to explain. *What do we learn beliefs are when we learn how to use the words*

---

[5] This paragraph corrects a misrepresentation of both Fodor's and Ryle's positions in my critical notice of Fodor's book in *Mind*, 1977, reprinted in *Brainstorms*, pp. 90–108.

'believe' and 'belief'? The first point to make is that we do not really learn what beliefs are when we learn how to use these words.[6] Certainly no one *tells us* what beliefs are, or if someone does, or if we happen to speculate on the topic on our own, the answer we come to, wise or foolish, will figure only weakly in our habits of thought about what people believe. We learn to *use* folk psychology – as a vernacular social technology, a craft – but we don't learn it self-consciously as a theory – we learn no meta-theory with the theory – and in this regard our knowledge of folk psychology is like our knowledge of the grammar of our native tongue. This fact does not make our knowledge of folk psychology entirely unlike human knowledge of explicit academic theories, however; one could probably be a good practising chemist and yet find it embarrassingly difficult to produce a satisfactory textbook definition of a metal or an ion.

There are no introductory textbooks of folk psychology (although Ryle's *The Concept of Mind* might be pressed into service), but many explorations of the field have been undertaken by ordinary language philosophers (under slightly different intentions), and more recently by more theoretically minded philosophers of mind, and from all this work an account of folk psychology – part truism and the rest controversy – can be gleaned. What are beliefs? *Roughly*, folk psychology has it that *beliefs* are information-bearing states of people that arise from perceptions, and which, together with appropriately related *desires*, lead to intelligent *action*. That much is relatively uncontroversial, but does folk psychology also have it that non-human animals have beliefs? If so, what is the role of language in belief? Are beliefs constructed of parts? If so, what are the parts? Ideas? Concepts? Words? Pictures? Are beliefs like speech acts or maps or instruction manuals or sentences? Is it implicit in folk psychology that beliefs enter into causal relations, or that they don't? How do decisions and intentions intervene between belief–desire complexes and actions? Are beliefs introspectible, and if so, what authority do the believer's pronouncements have?

All these questions deserve answers, but one must bear in mind that there are different reasons for being interested in the details of folk psychology. One reason is that it exists as a phenomenon, like a religion or a language or a dress code, to be studied with the techniques and attitudes of anthropology. It may be a myth, but it is a myth we live in, so it is an 'important' phenomenon in nature. A different reason is that it seems to be a *true* theory, by and large, and hence is a candidate – like the folk physics of magnets and unlike the folk science of astrology – for

[6] I think it is just worth noting that philosophers' use of 'believe' as the standard and general ordinary language term is a considerable distortion. We *seldom* talk about what people *believe*; we talk about what they *think* and what they *know*.

incorporation into science. These different reasons generate different but overlapping investigations. The anthropological question should include in its account of folk psychology whatever folk actually include in their theory, however misguided, incoherent, gratuitous some of it may be.[7] The proto-scientific quest, on the other hand, as an attempt to prepare folk theory for subsequent incorporation into or reduction to the rest of science, should be critical, and should *eliminate* all that is false or ill-founded, however well-entrenched in popular doctrine. (Thales thought that lodestones had souls, we are told. Even if most people agreed, this would be something to eliminate from the folk physics of magnets prior to 'reduction'.) One way of distinguishing the good from the bad, the essential from the gratuitous, in folk theory is to see what must be included in the theory to account for whatever predictive or explanatory success it seems to have in ordinary use. In this way we can criticize as we analyse, and it is even open to us in the end to discard folk psychology if it turns out to be a bad theory, and with it the presumed theoretical entities named therein. If we discard folk psychology as a theory, we would have to replace it with another theory, which while it did violence to many ordinary intuitions would explain the predictive power of the residual folk craft.

We use folk psychology all the time, to explain and predict each other's behaviour; we attribute beliefs and desires to each other with confidence – and quite unself-consciously – and spend a substantial portion of our waking lives formulating the world – not excluding ourselves – in these terms. Folk psychology is about as pervasive a part of our second nature as is our folk physics of middle-sized objects. How good is folk psychology? If we concentrate on its weaknesses we will notice that we often are unable to make sense of particular bits of human behaviour (our own included) in terms of belief and desire, even in retrospect; we often cannot predict accurately or reliably what a person will do or when; we often can find no resources within the theory for settling disagreements about particular attributions of belief or desire. If we concentrate on its strengths we find first that there are large areas in which it is extraordinarily reliable in its predictive power. Every time we venture out on a highway, for example, we stake our lives on the reliability of our general expectations about the perceptual beliefs, normal desires and decision proclivities of the other motorists. Second, we find that it is a theory of great generative power and efficiency. For instance, watching a film with a highly original and

---

[7] If the anthropologist marks part of the catalogue of folk theory as false, as an inaccurate or unsound account of the folk craft, he may speak of *false consciousness* or *ideology*; the role of such false theory in constituting a feature of the anthropological phenomenon is not diminished by its falseness.

unstereotypical plot, we see the hero smile at the villain and we all swiftly and effortlessly arrive at the same complex theoretical diagnosis: 'Aha!' we conclude (but perhaps not consciously), 'he wants her to think he doesn't know she intends to defraud his brother!' Third, we find that even small children pick up facility with the theory at a time when they have a very limited experience of human activity from which to induce a theory. Fourth, we find that we all use folk psychology knowing next to nothing about what actually happens inside people's skulls. 'Use your head' we are told, and we know some people are brainier than others, but our capacity to use folk psychology is quite unaffected by ignorance about brain processes – or even by large-scale misinformation about brain processes.

As many philosophers have observed, a feature of folk psychology that sets it apart from both folk physics and the academic physical sciences is the fact that explanations of actions citing beliefs and desires normally not only describe the provenance of the actions, but at the same time defend them as reasonable under the circumstances. They are reason-giving explanations, which make an ineliminable allusion to the rationality of the agent. Primarily for this reason, but also because of the pattern of strengths and weaknesses just described, I suggest that folk psychology might best be viewed as a rationalistic calculus of interpretation and prediction – an idealizing, abstract, instrumentalistic interpretation-method that has evolved because it works, and works because we have evolved. We approach each other as *intentional systems*,[8] that is, as entities whose behaviour can be predicted by the method of attributing beliefs, desires and rational acumen according to the following rough and ready principles:[9]

(1) A system's beliefs are those it *ought to have*, given its perceptual capacities, its epistemic needs, and its biography. Thus, in general, its beliefs are both true and relevant to its life, and when false beliefs are attributed, special stories must be told to explain how the error resulted from the presence of features in the environment that are deceptive relative to the perceptual capacities of the system.

(2) A system's desires are those it *ought to have*, given its biological needs and the most practicable means of satisfying them. Thus intentional systems desire survival and procreation, and hence desire food, security, health, sex, wealth, power, influence, and so forth, and also whatever local arrangements tend (in their eyes – given their

[8] See my 'Intentional Systems' (1971).
[9] For a more elaborate version of similar principles, see Lewis 1974.

beliefs) to further these ends in appropriate measure. Again, 'abnormal' desires are attributable if special stories can be told.

(3) A system's behaviour will consist of those acts that *it would be rational* for an agent with those beliefs and desires to perform.

In (1) and (2) 'ought to have' means 'would have if it were *ideally* ensconced in its environmental niche'. Thus all dangers and vicissitudes in its environment it will *recognize as such* (i.e. *believe* to be dangers) and all the benefits – relative to its needs, of course – it will *desire*. When a fact about its surroundings is particularly relevant to its current projects (which themselves will be the projects such a being ought to have in order to get ahead in its world) it will *know* that fact, and act accordingly. And so forth and so on. This gives us the notion of an ideal epistemic and conative operator or agent, relativized to a set of needs for survival and procreation and to the environment(s) in which its ancestors have evolved and to which it is adapted. But this notion is still too crude and overstated. For instance, a being may come to have an epistemic need that its perceptual apparatus cannot provide for (suddenly all the green food is poisonous but alas it is colourblind), hence the relativity to perceptual capacities. Moreover, it may or may not have had the occasion to learn from experience about something, so its beliefs are also relative to its biography in this way: it will have learned what it ought to have learned, *viz.* what it had been given evidence for in a form compatible with its cognitive apparatus – providing the evidence was 'relevant' to its project then.

But this is still too crude, for we understand that evolution does not give us a best of all possible worlds, but only a passable jury-rig, so we should look for design shortcuts that in specifiably abnormal circumstances yield false perceptual beliefs, etc. (We are not immune to illusions – which we would be if our perceptual systems were *perfect*.) To offset the design shortcuts we should also expect design bonuses: circumstances in which the 'cheap' way for nature to design a cognitive system has the side benefit of giving good, reliable results even outside the environment in which the system evolved. Our eyes are well adapted for giving us true beliefs on Mars as well as on Earth – because the cheap solution for our Earth-evolving eyes happens to be a more general solution.[10]

I propose that we can continue the mode of thinking just illustrated *all the way in* – not just for eye-design, but for deliberation-design and belief-design and strategy-concocter-design. In using this optimistic set of assumptions (nature has built us to do things right; look for systems to believe the truth and love the good) we impute no occult

---

[10] Cf. Sober (unpublished) for useful pioneering exploration of these topics.

powers to epistemic needs, perceptual capacities and biography, but only the powers common sense already imputes to evolution and learning.

In short, we treat each other as if we were rational agents, and this myth – for surely we are not all that rational – works very well because we are *pretty* rational. This single assumption, in combination with home truths about our needs, capacities and typical circumstances, generates both an intentional interpretation of us as believers and desirers and actual predictions of behaviour in great profusion. I am claiming, then, that folk psychology can best be viewed as a sort of logical behaviourism: *what it means* to say that someone believes that *p*, is that that person is disposed to behave in certain ways under certain conditions. What ways under what conditions? The ways it would be rational to behave, given the person's other beliefs and desires. The answer looks in danger of being circular, but consider: an account of what it is for an element to have a particular valence will similarly make ineliminable reference to the valences of other elements. What one is given with valence-talk is a whole system of interlocking attributions, which is saved from vacuity by yielding independently testable predictions.

I have just described in outline a *method* of predicting and explaining the behaviour of people and other intelligent creatures. Let me distinguish two questions about it: (1) is it something we could do and (2) is it something we in fact do? I think the answer to (1) is obviously yes, which is not to say the method will always yield good results. That much one can ascertain by reflection and thought experiment. Moreover, one can recognize that the method is familiar. Although we don't usually use the method self-consciously, we do use it self-consciously on those occasions when we are perplexed by a person's behaviour, and then it often yields satisfactory results. Moreover, the ease and naturalness with which we resort to this self-conscious and deliberate form of problem-solving provide some support for the claim that what we are doing on those occasions is not *switching methods* but simply becoming self-conscious and explicit about what we ordinarily accomplish tacitly or unconsciously.

No other view of folk psychology, I think, can explain the fact that we do so well predicting each other's behaviour on such slender and peripheral evidence; treating each other as intentional systems works (to the extent that it does) because we really are well designed by evolution and hence we *approximate* to the ideal version of ourselves exploited to yield the predictions. But not only does evolution not guarantee that we will always do what is rational; it guarantees that we won't. If we are designed by evolution, then we are almost certainly nothing more than a bag of tricks, patched together by a *satisficing*[11] Nature, and no better than

---
[11] The term is Herbert Simon's (e.g. 1969).

our ancestors had to be to get by. Moreover, the demands of nature and the demands of a logic course are not the same. Sometimes – even *normally* in certain circumstances – it pays to jump to conclusions swiftly (and even to forget that you've done so), so by most philosophical measures of rationality (logical consistency, refraining from invalid inference) there has probably been some positive evolutionary pressure in favour of 'irrational' methods.[12]

How rational are we? Recent research in social and cognitive psychology suggests we are *minimally* rational, appallingly ready to leap to conclusions or be swayed by logically irrelevant features of situations,[13] but this jaundiced view is an illusion engendered by the fact that these psychologists are deliberately trying to produce situations that provoke irrational responses – inducing pathology in a system by putting strain on it – and succeeding, being good psychologists. No one would hire a psychologist to prove that people will choose a paid vacation to a week in jail if offered an informed choice. At least not in the better psychology departments. A more optimistic impression of our rationality is engendered by a review of the difficulties encountered in artificial intelligence research. Even the most sophisticated AI programmes stumble blindly into misinterpretations and misunderstandings that even small children reliably evade without a second thought.[14] From this vantage point we seem marvellously rational.

However rational we are, it is the myth of our rational agenthood that structures and organizes our attributions of belief and desire to others, and that regulates our own deliberations and investigations. We aspire to rationality, and without the myth of our rationality the concepts of belief and desire would be uprooted. Folk psychology, then, is *idealized* in that it produces its predictions and explanations by calculating in a normative

---

[12] While in general true beliefs have to be more useful than false beliefs (and hence a system ought to have true beliefs), in special circumstances it may be better to have a few false beliefs. For instance it might be better for beast B to have some false beliefs about whom B can beat up and whom B can't. Ranking B's likely antagonists from ferocious to pushover, we certainly want B to believe it can't beat up all the ferocious ones, and can beat up all the obvious pushovers, but it is better (because it 'costs less' in discrimination tasks and protects against random perturbations such as bad days and lucky blows) for B to extend 'I can't beat up x' to cover even some beasts it can in fact beat up. *Erring on the side of prudence* is a well recognized good strategy, and so Nature can be expected to have valued it on occasion when it came up. An alternative strategy in this instance would be to abide by the rule: avoid conflict with penumbral cases. But one might have to 'pay more' to implement that strategy than to implement the strategy designed to produce, and rely on, some false beliefs.

[13] See, e.g. Tversky and Kahneman 1974; and Nisbett and Ross 1978.

[14] Roger Schank's (1977; Schank and Abelson 1977) efforts to get a computer to 'understand' simple but normally gappy stories is a good illustration.

system; it predicts what we *will* believe, desire, and do, by determining what we *ought* to believe, desire, and do.[15]

Folk psychology is *abstract* in that the beliefs and desires it attributes are not – or need not be – presumed to be intervening distinguishable states of an internal behaviour-causing system. (The point will be enlarged upon later.) The role of the concept of belief is like the role of the concept of a centre of gravity, and the calculations that yield the predictions are more like the calculations one performs with a parallelogram of forces than like the calculations one performs with a blueprint of internal levers and cogs.

Folk psychology is thus *instrumentalistic* in a way the most ardent realist should permit: people really do have beliefs and desires, on my version of folk psychology, just the way they really have centres of gravity and the earth has an Equator.[16] Reichenbach distinguished between two sorts of referents for theoretical terms: *illata* – posited theoretical entities – and *abstracta* – calculation-bound entities or logical constructs.[17] Beliefs and desires of folk psychology (but not all mental events and states) are *abstracta*.

This view of folk psychology emerges more clearly in contrast to a diametrically opposed view, each of whose tenets has been held by some philosopher, and at least most of which have been espoused by Fodor:

Beliefs and desires, just like pains, thoughts, sensations and other episodes, are taken by folk psychology to be real, intervening, internal states or events, in causal interaction, subsumed under covering laws of causal stripe. Folk psychology is not an idealized, rationalistic calculus but a naturalistic, empirical, descriptive theory, imputing causal regularities discovered by extensive induction over experience. To suppose two people share a belief is to suppose them to be ultimately in some structurally similar internal condition, e.g. for them to have the same words of Mentalese written in the functionally relevant places in their brains.

I want to deflect this head-on collision of analyses by taking two steps. First, I am prepared to grant a measure of the claims made by the opposition. *Of course* we don't all sit in the dark in our studies like mad

---

[15] It tests its predictions in two ways: action predictions it tests directly by looking to see what the agent does; belief and desire predictions are tested indirectly by employing the predicted attributions in further predictions of eventual action. As usual, the Duhemian thesis holds: belief and desire attributions are under-determined by the available data.

[16] Michael Friedman's 'Theoretical Explanation' (in this volume) provides an excellent analysis of the role of instrumentalistic thinking within realistic science. Scheffler (1963) provides a useful distinction between *instrumentalism* and *fictionalism*. In his terms I am characterizing folk psychology as instrumentalistic, not fictionalistic.

[17] Reichenbach 1938: 211–12. 'Our observations of concrete things confer a certain probability on the existence of *illata* – nothing more . . . Second, there are inferences to *abstracta*. These inferences are . . . equivalences, not probability inferences. Consequently, the existence of abstracta is reducible to the existence of concreta. There is, therefore, no problem of their objective existence; their status depends on a convention.'

Leibnizians rationalistically excogitating behavioural predictions from pure, idealized concepts of our neighbours, nor do we derive all our readiness to attribute desires from a careful generation of them from the ultimate goal of survival. We may observe that some folks seem to desire cigarettes, or pain, or notoriety (we observe this by hearing them tell us, seeing what they choose, etc.) and without any conviction that these people, given their circumstances, ought to have these desires, we attribute them anyway. So rationalistic generation of attributions is augmented and even corrected on occasion by empirical generalizations about belief and desire that guide our attributions and are learned more or less inductively. For instance, small children believe in Santa Claus, people are inclined to believe the more self-serving of two interpretations of an event in which they are involved (unless they are depressed), and people can be made to want things they don't need by making them believe that glamorous people like those things. And so forth in familiar profusion. This folklore does not consist in *laws* – even probabilistic laws – but some of it is being turned into science of a sort, e.g. theories of 'hot cognition' and cognitive dissonance. I grant the existence of all this naturalistic generalization, and its role in the normal calculations of folk psychologists – i.e. all of us. People do rely on their own parochial group of neighbours when framing intentional interpretations. That is why people have so much difficulty understanding foreigners – their behaviour, to say nothing of their languages. They impute more of their own beliefs and desires, and those of their neighbours, than they would if they followed my principles of attribution slavishly. Of course this is a perfectly reasonable shortcut for people to take, even when it often leads to bad results. We are in this matter, as in most, satisficers, not optimizers, when it comes to information gathering and theory construction. I would insist, however, that all this empirically obtained lore is laid over a fundamental generative and normative framework that has the features I have described.

My second step away from the conflict I have set up is to recall that the issue is not what folk psychology as found in the field truly is, but what it is at its best, what deserves to be taken seriously and incorporated into science. It is not particularly to the point to argue against me that folk psychology is *in fact* committed to beliefs and desires as distinguishable, causally interacting *illata*; what must be shown is that it ought to be. The latter claim I will deal with in due course. The former claim I *could* concede without embarrassment to my overall project, but I do not concede it, for it seems to me that the evidence is quite strong that our ordinary notion of belief has next to nothing of the concrete in it. Jacques shoots his uncle dead in Trafalgar Square and is apprehended on the spot by Sherlock; Tom reads about it in the *Guardian* and Boris learns of it in

Pravda. Now Jacques, Sherlock, Tom and Boris have had remarkably *different* experiences – to say nothing of their earlier biographies and future prospects – but there is one thing they share: they all believe that a Frenchman has committed murder in Trafalgar Square. They did not all *say* this, not even 'to themselves'; *that proposition* did not, we can suppose, 'occur to' any of them, and even if it had, it would have had entirely different import for Jacques, Sherlock, Tom and Boris. Yet they all believe that a Frenchman committed murder in Trafalgar Square. This is a shared property that is, as it were, visible only from one very limited point of view – the point of view of folk psychology. Ordinary folk psychologists have no difficulty imputing such useful but elusive commonalities to people. If they then insist that in doing so they are postulating a similarly structured object, as it were, in each head, this is a gratuitous bit of misplaced concreteness, a regrettable lapse in ideology.

But in any case there is no doubt that folk psychology is a mixed bag, like folk productions generally, and there is no reason in the end not to grant that it is much more complex, variegated (and in danger of incoherence) than my sketch has made it out to be. The *ordinary* notion of belief no doubt does place beliefs somewhere midway between being *illata* and being *abstracta*. What this suggests to me is that the concept of belief found in ordinary understanding, i.e. in folk psychology, is unappealing as a scientific concept. I am reminded of Anaxagoras' strange precursor to atomism: the theory of seeds. There is a portion of everything in everything, he is reputed to have claimed. Every object consists of an infinity of seeds, of all possible varieties. How do you make bread out of flour, yeast and water? Flour contains bread seeds in abundance (but flour seeds predominate – that's what makes it flour), and so do yeast and water, and when these ingredients are mixed together, the bread seeds form a new majority, so bread is what you get. Bread nourishes by containing flesh and blood and bone seeds in addition to its majority of bread seeds. Not good theoretical entities, these seeds, for as a sort of bastardized cross between properties and proper parts they have a penchant for generating vicious regresses, and their identity conditions are problematic to say the least.

Beliefs are rather like that. There seems no comfortable way of avoiding the claim that we have an infinity of beliefs, and common intuition does not give us a stable answer to such puzzles as whether the belief that 3 is greater than 2 is none other than the belief that 2 is less than 3. The obvious response to the challenge of an infinity of beliefs with slippery identity conditions is to suppose these beliefs are not all 'stored separately'; many – in fact *most* if we are really talking about infinity – will be stored *implicitly* in virtue of the *explicit* storage of a few (or a few million)

– the *core beliefs*.[18] The core beliefs will be 'stored separately', and they look like promising *illata* in contrast to the *virtual* or *implicit* beliefs which look like paradigmatic *abstracta*. But although this might turn out to be the way our brains are organized, I suspect things will be more complicated than this: there is no reason to suppose the core *elements*, the concrete, salient, separately stored representation-tokens (and there must be some such elements in any complex information processing system), will explicitly represent (or *be*) a subset of our *beliefs* at all. That is, if you were to sit down and write out a list of a thousand or so of your paradigmatic beliefs, *all* of them could turn out to be virtual, only implicitly stored or represented, and what was explicitly stored would be information (e.g. about memory addresses, procedures for problem-solving, or recognition, etc.) that was entirely unfamiliar. It would be folly to prejudge this empirical issue by insisting that our core representations of information (whichever they turn out to be) are beliefs *par excellence*, for when the facts are in our intuitions may instead support the contrary view: the least controversial self-attributions of belief may pick out beliefs that from the vantage point of developed cognitive theory are invariably virtual.[19]

In such an eventuality what could we say about the *causal* roles we assign ordinarily to beliefs (e.g. 'Her belief that John knew her secret caused her to blush')? We could say that whatever the core elements were in virtue of which she virtually believed that John knew her secret, they, the core elements, played a direct causal role (somehow) in triggering the blushing response. We would be wise, as this example shows, not to tamper with our *ordinary* catalogue of beliefs (virtual though they might all turn out to be), for these are predictable, readily understandable, manipulable regularities in psychological phenomena in spite of their apparent neutrality with regard to the explicit/implicit (or core/virtual) distinction. What Jacques, Sherlock, Boris and Tom have in common is probably only a virtual belief 'derived' from largely different explicit stores of information in each of them, but virtual or not, it is their sharing of *this* belief that would explain (or permit us to predict) in some imagined circumstances their all taking the same action when given the same new information. ('And now for one million dollars, Tom [Jacques, Sherlock, Boris], answer our jackpot question correctly: has a French citizen ever committed a major crime in London?')

At the same time we want to cling to the equally ordinary notion that beliefs can cause not only actions, but blushes, verbal slips, heart attacks and the like. Much of the debate over whether or not intentional explanations are causal explanations can be bypassed by noting how the

[18] See my 'Brain Writing and Mind Reading', 1975. See also Fodor 1975, and Field 1978.
[19] See Field 1978: 55, n. 12 on 'minor concessions' to such instrumentalistic treatments of belief.

core elements, *whatever they may be*, can be cited as playing the causal role, while belief remains virtual. 'Had Tom not believed that p and wanted that q, he would not have done A.' Is this a causal explanation? It is tantamount to this: Tom was in some one of an indefinitely large number of structurally different states of type B that have in common just that each one of them licenses attribution of belief that p and desire that q in virtue of its normal relations with many other states of Tom, and this state, whichever one it was, was causally sufficient, given the 'background conditions' of course, to initiate the intention to perform A, and thereupon A was performed, and had he not been in one of those indefinitely many type B states, he would not have done A. One can call this a causal explanation because it talks about causes, but it is surely as unspecific and unhelpful as a causal explanation can get. It commits itself to there being some causal explanation or other falling within a very broad area (i.e. the intentional interpretation is held to be supervenient on Tom's bodily condition), but its true informativeness and utility in actual prediction lie, not surprisingly, in its assertion that Tom, however his body is currently structured, has a particular set of these elusive intentional properties, beliefs and desires.

The ordinary notion of belief is pulled in two directions. If we want to have *good* theoretical entities, good *illata*, or good logical constructs, good *abstracta*, we will have to jettison some of the ordinary freight of the concepts of belief and desire. So I propose a divorce. Since we seem to have both notions wedded in folk psychology, let's split them apart and create two new theories: one strictly abstract, idealizing, holistic, in-strumentalistic – pure intentional system theory – and the other a concrete, micro-theoretical science of the actual realization of those intentional systems – what I will call sub-personal cognitive psychology. By exploring their differences and interrelations, we should be able to tell whether any plausible 'reductions' are in the offing.

2

The first new theory, intentional system theory, is envisaged as a close kin of – and overlapping with – such already existing disciplines as decision theory and game theory, which are similarly abstract, normative and couched in intentional language. It borrows the ordinary terms, 'belief' and 'desire' but gives them a technical meaning within the theory. It is a sort of holistic logical behaviourism because it deals with the prediction and explanation from belief–desire profiles of the actions of whole systems (either alone in environments or in interaction with other intentional systems), but treats the individual realizations of the systems as black boxes. The *subject* of all the intentional attributions is the whole system (the person, the animal, or even the corporation or nation)[20] rather than any of its parts, and individual beliefs and desires are not attributable in isolation, independently of other belief and desire attributions. The latter point distinguishes intentional system theory most clearly from Ryle's logical behaviourism, which took on the impossible burden of characterizing individual beliefs (and other mental states) as particular individual dispositions to outward behaviour.

The theory deals with the 'production' of new beliefs and desires from old, *via* an interaction among old beliefs and desires, features in the environment, and the system's actions, and this creates the illusion that the theory contains naturalistic descriptions of internal processing in the systems the theory is about, when in fact the processing is all in the manipulation of the theory, and consists in updating the intentional characterization of the whole system according to the rules of attribution. An analogous illusion of process would befall a naive student who, when confronted with a parallelogram of forces, supposed that it pictured a mechanical linkage of rods and pivots of some kind instead of being simply a graphic way of representing and plotting the effect of several simultaneously acting forces.

Richard Jeffrey (1970), in developing his concept of probability kinematics, has usefully drawn attention to an analogy with the distinction in physics between kinematics and dynamics. In kinematics,

you talk about the propagation of motions throughout a system in terms of such constraints as rigidity and manner of linkage. It is the physics of position and time, in terms of which you can talk about velocity and acceleration, but not about force and mass. When you talk about forces – *causes* of accelerations – you are in the realm of dynamics (172).

Kinematics provides a simplified and idealized level of abstraction appropriate for many purposes – e.g. for the *initial* design development of a gearbox – but when one must deal with more concrete details of systems – e.g. when the gearbox designer must worry about friction, bending, energetic efficiency and the like – one must switch to dynamics for more detailed and reliable predictions, at the cost of increased complexity and diminished generality. Similarly one can approach the study of belief (and desire and so forth) at a highly abstract level, ignoring problems of realization and simply setting out what the normative demands on the design of a believer are. For instance, one can ask such questions as 'What must a system's epistemic capabilities and propensities be for it to survive in environment A?'[21] or 'What must this system already know in order for

[20] See my 'Conditions of Personhood' (1976).
[21] Cf. Campbell 1973, and his William James lectures (Harvard U.P., forthcoming).

it to be able to learn B?' or 'What intentions must this system have in order to mean something by saying something?'[22]

Intentional system theory deals just with the performance specifications of believers while remaining silent on how the systems are to be implemented. In fact this neutrality with regard to implementation is the most useful feature of intentional characterizations. Consider, for instance, the role of intentional characterizations in evolutionary biology. If we are to explain the evolution of complex behavioural capabilities or cognitive talents by natural selection, we must note that it is the intentionally characterized capacity (e.g. the capacity to acquire a belief, a desire, to perform an intentional action) that has survival value, however it happens to be realized as a result of mutation. If a particularly noxious insect makes its appearance in an environment, the birds and bats with a survival advantage will be those that come to believe this insect is not good to eat. In view of the vast differences in neural structure, genetic background and perceptual capacity between birds and bats, it is highly unlikely that this useful trait they may come to share has a common description at any level more concrete or less abstract than intentional system theory. It is not only that the intentional predicate is a projectible predicate in evolutionary theory; since it is more general than its species-specific counterpart predicates (which characterize the successful mutation just in birds, or just in bats), it is preferable. So from the point of view of evolutionary biology, we would not want to 'reduce' all intentional characterizations even if we knew in particular instances what the physiological implementation was.

This level of generality is essential if we want a theory to have anything meaningful and defensible to say about such topics as intelligence in general (as opposed, say, to just human or even terrestrial or natural intelligence), or such grand topics as meaning or reference or representation. Suppose, to pursue a familiar philosophical theme, we are invaded by Martians, and the question arises: do they have beliefs and desires? Are they that much *like us*? According to intentional system theory, if these Martians are smart enough to get here, then they most certainly have beliefs and desires – in the technical sense proprietary to the theory – no matter what their internal structure, and no matter how our folk-psychological intuitions rebel at the thought.

This principled blindness of intentional system theory to internal structure seems to invite the retort:[23] but there has to be *some* explanation of the *success* of intentional prediction of the behaviour of systems. It isn't

---

[22] The questions of this variety are familiar, of course, to philosophers, but are now becoming equally familiar to researchers in artificial intelligence.

[23] From Ned Block and Jerry Fodor, *inter alia*, in conversation.

just magic. It isn't a mere coincidence that one can generate all these *abstracta*, manipulate them *via* some version of practical reasoning, and come up with an action prediction that has a good chance of being true. There must be some way in which the internal processes of the system mirror the complexities of the intentional interpretation, or its success would be a miracle.

Of course. This is all quite true and important. Nothing without a great deal of structural and processing complexity could conceivably realize an intentional system of any interest, and the complexity of the realization will surely bear a striking resemblance to the complexity of the in- strumentalistic interpretation. Similarly, the success of valence theory in chemistry is no coincidence, and people were entirely right to expect that deep microphysical similarities would be discovered between elements with the same valence, and that the structural similarities found would explain the dispositional similarities. But since people and animals are unlike atoms and molecules not only in being the products of a complex evolutionary history, but also in being the products of their individual learning histories, there is no reason to suppose that individual (human) believers that p – like individual (carbon) atoms with valence 4 – regulate their dispositions with *exactly* the same machinery. Discovering the constraints on design and implementation variation, and demonstrating how particular species and individuals in fact succeed in realizing intentional systems is the job for the third theory: sub-personal cognitive psychology.

3

The task of sub-personal cognitive psychology is to explain something that at first glance seems utterly mysterious and inexplicable. The brain, as intentional system theory and evolutionary biology show us, is a *semantic engine*; its task is to discover what its multifarious inputs *mean*, to discriminate them by their significance and 'act accordingly'.[24] That's what brains *are for*. But the brain, as physiology or plain common sense shows us, is just a *syntactic engine*; all it can do is discriminate its inputs by their structural, temporal, and physical features, and let its entirely mechanical activities be governed by these 'syntactic' features of its

---

[24] More accurately if less picturesquely, the brain's task is to come to produce internal mediating responses that reliably vary in concert with variation in the actual environmen- tal significance (the natural and non-natural meanings, in Grice's (1957) sense) of their distal causes and independently of meaning–irrelevant variations in their proximal causes, and moreover to respond to its own mediating responses in ways that systematically tend to improve the creature's prospects in its environment if the mediating responses are varying as they ought to vary.

inputs. That's all brains *can do*. Now how does the brain manage to get semantics from syntax? How could *any* entity (how could a genius, or an angel, or God) get the semantics of a system from nothing but its syntax? It couldn't. The syntax of a system doesn't determine its semantics. By what alchemy, then, does the brain extract semantically reliable results from syntactically driven operations? It cannot be designed to do an impossible task, but it could be designed to *approximate* the impossible task, to *mimic* the behaviour of the impossible object (the semantic engine) by capitalizing on close (close enough) fortuitous correspondences between structural regularities – of the environment and of its own internal states and operations – and semantic types.

The basic idea is familiar. An animal needs to know when it has satisfied the goal of finding and ingesting food, but it settles for a friction-in-the-throat-followed-by-stretched-stomach detector, a mechanical switch turned on by a relatively simple mechanical condition that *normally* co-occurs with the satisfaction of the animal's 'real' goal. It's not fancy, and can easily be exploited to trick the animal into either eating when it shouldn't or leaving off eating when it shouldn't, but it does well enough by the animal in its normal environment. Or suppose I am monitoring telegraph transmissions and have been asked to intercept all *death threats* (but only death threats in English – to make it 'easy'). I'd like to build a machine to save me the trouble of interpreting semantically every message sent, but how could this be done? No machine could be designed to do the job perfectly, for that would require defining the semantic category *death threat in English* as some tremendously complex feature of strings of alphabetic symbols, and there is utterly no reason to suppose this could be done in a principled way. (If somehow by brute-force inspection and subsequent enumeration we could list all and only the English death threats of, say, less than a thousand characters, we could easily enough build a filter to detect them, but we are looking for a principled, projectible, extendable method.) A really crude device could be made to discriminate all messages containing the symbol strings

... I will kill you ...

or

... you ... die ... unless ...

or

... (for some finite disjunction of likely patterns to be found in English death threats).

This device would have some utility, and further refinements could screen the material that passed this first filter, and so on. An unpromising beginning for constructing a sentence understander, but if you want to get semantics out of syntax (whether the syntax of messages in a natural language or the syntax of afferent neuron impulses), variations on this basic strategy are your only hope.[25] You must put together a bag of tricks and hope nature will be kind enough to let your device get by. Of course some tricks are elegant, and appeal to deep principles of organization, but in the end all one can hope to produce (all natural selection can have produced) are systems that *seem* to discriminate meanings by actually discriminating things (tokens of no doubt wildly disjunctive types) that co-vary reliably with meanings.[26] Evolution has designed our brains not only to do this but to evolve and follow strategies of self-improvement in this activity during their individual lifetimes.[27]

It is the task of sub-personal cognitive psychology to propose and test models of such activity – of pattern recognition or stimulus generalization, concept learning, expectation, learning, goal-directed behaviour, problem-solving – that not only produce a simulacrum of genuine content-sensitivity, but that do this in ways demonstrably like the way people's brains do it, exhibiting the same powers and the same vulnerabilities to deception, overload and confusion. It is here that we will find our good theoretical entities, our useful *illata*, and while some of them may well resemble the familiar entities of folk psychology – beliefs, desires, judgments, decisions – many will certainly not.[28] The only similarity we can be sure of discovering in the *illata* of sub-personal cognitive psychology is the

25 One might think that while *in principle* one cannot derive the semantics of a system from nothing but its syntax, *in practice* one might be able to cheat a little and exploit syntactic features that don't *imply* a semantical interpretation, but strongly suggest one. For instance, faced with the task of deciphering isolated documents in an entirely unknown and alien language, one might note that while the symbol that *looks like* a duck doesn't *have* to mean 'duck', there is a good chance that it does, especially if the symbol that looks like a wolf seems to be eating the symbol that looks like a duck, and not *vice versa*. Call this *hoping for hieroglyphics* and note the form it has taken in psychological theories from Locke to the present: we will be able to tell which mental representations are which (which idea is the idea of *dog* and which of *cat*) because the former will look like a dog and the latter like a cat. This is all very well as a crutch for us observers on the outside, trying to assign content to the events in some brain, but it is of no use to the brain . . . because brains don't know what dogs look like! Or better, this cannot be the brain's fundamental method of eking semantic classes out of raw syntax, for any brain (or brain part) that could be said – in an extended sense – to know what dogs look like would be a brain (or brain part) that had already solved its problem, that was already (a simulacrum of) a semantic engine. But this is still misleading, for brains in any event do not *assign* content to their own events in the way observers might: brains *fix* the content of their internal events in the act of reacting as they do. There are good reasons for positing *mental images* of one sort or another in cognitive theories (see 'Two Approaches to Mental Images' in *Brainstorms* pp. 174–89) but hoping for hieroglyphics isn't one of them, though I suspect it is covertly influential.

26 I take this point to be closely related to Davidson's reasons for claiming there can be no psycho-physical laws, but I am unsure that Davidson wants to draw the same conclusions from it that I do. See Davidson 1970.

27 This claim is defended in my 'Why the law of effect will not go away' (1974).

28 See, for instance, Stephen Stich's (1978) concept of subdoxastic states.

intentionality of their labels.[29] They will be characterized as events with content, bearing information, signalling this and ordering that.

In order to give the *illata* these labels, in order to maintain any intentional interpretation of their operation at all, the theorist must always keep glancing outside the system, to see what normally produces the configuration he is describing, what effects the system's responses normally have on the environment, and what benefit normally accrues to the whole system from this activity. In other words the cognitive psychologist cannot ignore the fact that it is the realization of an intentional system he is studying on pain of abandoning semantic interpretation and hence psychology. On the other hand, progress in sub-personal cognitive psychology will blur the boundaries between it and intentional system theory, knitting them together much as chemistry and physics have been knit together.

The alternative of ignoring the external world and its relations to the internal machinery (what Putnam has called psychology in the narrow sense, or methodological solipsism, and Keith Gunderson lampoons as black world glass box perspectivalism)[30] is not really psychology at all,

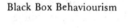

Black Box Behaviourism   Black World Glass Box Perspectivalism

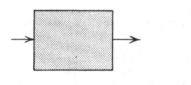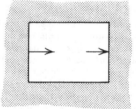

but just at best abstract neurophysiology – pure internal syntax with no hope of a semantic interpretation. Psychology 'reduced' to neurophysiology in this fashion would not be psychology, for it would not be able to provide an explanation of the regularities it is psychology's particular job to explain: the reliability with which 'intelligent' organisms can cope with their environments and thus prolong their lives. Psychology can, and should, work towards an account of the physiological foundations of psychological processes, not by eliminating psychological or intentional characterizations of those processes, but by exhibiting how the brain

[29] See my 'Reply to Arbib and Gunderson', in *Brainstorms*, pp. 23–38.
[30] In his reply to Fodor's 'Methodological Solipsism as a Research Strategy in Psychology' at the Cincinnati Colloquium on Philosophy of Psychology, February 1978.

implements the intentionally characterized performance specifications of sub-personal theories.[31]

Friedman, discussing the current perplexity in cognitive psychology, suggests that the problem

is the direction of reduction. Contemporary psychology tries to explain *individual* cognitive activity independently from *social* cognitive activity, and then tries to give a *micro* reduction of social cognitive activity – that is, the use of a public language – in terms of a prior theory of individual cognitive activity. The opposing suggestion is that we first look for a theory of social activity, and then try to give a *macro* reduction of individual cognitive activity – the activity of applying concepts, making judgments, and so forth – in terms of our prior social theory.[32]

With the idea of macro-reduction in psychology I largely agree, except that Friedman's identification of the macro level as explicitly *social* is only part of the story. The cognitive capacities of non-language-using animals (and Robinson Crusoes, if there are any) must also be accounted for, and not just in terms of an analogy with the practices of us language users. The macro level *up* to which we should relate micro-processes in the brain in order to understand them as psychological is more broadly the level of organism–environment interaction, development and evolution. That level includes social interaction as a particularly important part,[33] but still a proper part.

There is no way to capture the semantic properties of things (word tokens, diagrams, nerve impulses, brain states) by a micro-reduction. Semantic properties are not just relational but, you might say, super-relational, for the relation a particular vehicle of content, or token, must bear in order to have content is not just a relation it bears to other similar things (e.g. other tokens, or parts of tokens, or sets of tokens, or causes of tokens) but a relation between the token and the whole life – and counter-factual life[34] – of the organism it 'serves' *and* that organism's requirements for survival *and* its evolutionary ancestry.

4

Of our three psychologies – folk psychology, intentional system theory, and sub-personal cognitive psychology – what then might reduce to what?

---

[31] I treat methodological solipsism in (much) more detail in 'Beyond Belief', in Andrew Woodfield, ed. *Thought and Object.*

[32] Michael Friedman, 'Theoretical Explanation', this volume, pp. 15–16.

[33] See Tyler Burge 1979.

[34] What I mean is this: counterfactuals enter because content is in part a matter of the *normal* or *designed* role of a vehicle whether or not it ever gets to play that role. Cf. Sober (unpublished).

Certainly the one-step micro-reduction of folk psychology to physiology alluded to in the slogans of the early identity theorists will never be found – and should never be missed, even by staunch friends of materialism and scientific unity. A prospect worth exploring, though, is that folk psychology (more precisely, the part of folk psychology worth caring about) reduces – conceptually – to intentional system theory. What this would amount to can best be brought out by contrasting this proposed conceptual reduction with more familiar alternatives: 'type-type identity theory' and 'Turing machine functionalism'. According to type-type identity theory, for every mentalistic term or predicate 'M', there is some predicate 'P' *expressible in the vocabulary of the physical sciences* such that a creature is M if and only if it is P. In symbols:

(1) $(x)(Mx \equiv Px)$

This is reductionism with a vengeance, taking on the burden of replacing, in principle, all mentalistic predicates with co-extensive predicates composed truth-functionally from the predicates of physics. It is now widely agreed to be hopelessly too strong a demand. Believing that cats eat fish is, intuitively, a *functional* state that might be variously implemented physically, so there is no reason to suppose the commonality referred to on the left-hand side of (1) can be reliably picked out by any predicate, however complex, of physics. What is needed to express the predicate on the right-hand side is, it seems, a physically neutral language for speaking of functions and functional states, and the obvious candidates are the languages used to describe automata – for instance, Turing machine language.

The Turing machine functionalist then proposes

(2) $(x)(Mx \equiv x$ realizes some Turing machine k in logical state A)

In other words, for two things both to believe that cats eat fish they need not be physically similar in any specifiable way, but they must both be in a 'functional' condition specifiable in principle in the most general functional language; they must share a Turing machine description according to which they are both in some particular logical state. This is still a reductionist doctrine, for it proposes to identify each mental type with a functional type picked out in the language of automata theory. But this is still too strong, for there is no more reason to suppose Jacques, Sherlock, Boris and Tom 'have the same programme' in *any* relaxed and abstract sense, considering the differences in their nature and nurture, than that their brains have some crucially identical physico-chemical feature. We must weaken the requirements for the right-hand side of our formula still further.

Consider

(3) (x)(x believes that p ≡ x can be predictively attributed the belief that p)

This appears to be blatantly circular and uninformative, with the language on the right simply mirroring the language on the left. But all we need to make an informative answer of this formula is a systematic way of making the attributions alluded to on the right-hand side. Consider the parallel case of Turing machines. What do two different realizations or embodiments of a Turing machine have in common when they are in the same logical state? Just this: there is a system of description such that according to it both are described as being realizations of some particular Turing machine, and according to this description, which is predictive of the operation of both entities, both are in the same state of that Turing machine's machine table. One doesn't *reduce* Turing machine talk to some more fundamental idiom; one *legitimizes* Turing machine talk by providing it with rules of attribution and exhibiting its predictive powers. If we can similarly legitimize 'mentalistic' talk, we will have no need of a reduction, and that is the point of the concept of an intentional system. Intentional systems are supposed to play a role in the legitimization of mentalistic predicates parallel to the role played by the abstract notion of a Turing machine in setting down rules for the interpretation of artifacts as computational automata. I fear my concept is woefully informal and unsystematic compared with Turing's, but then the domain it attempts to systematize – our everyday attributions in mentalistic or intentional language – is itself something of a mess, at least compared with the clearly defined field of recursive function theory, the domain of Turing machines.

The analogy between the theoretical roles of Turing machines and intentional systems is more than superficial. Consider that warhorse in the philosophy of mind, Brentano's Thesis that intentionality is the mark of the mental: all mental phenomena exhibit intentionality and no physical phenomena exhibit intentionality. This has been traditionally taken to be an *irreducibility* thesis: the mental, in virtue of its intentionality, cannot be reduced to the physical. But given the concept of an intentional system, we can construe the first half of Brentano's Thesis – all mental phenomena are intentional – as a *reductionist* thesis of sorts, parallel to Church's Thesis in the foundation of mathematics.

According to Church's Thesis, every 'effective' procedure in mathematics is recursive, that is, Turing-computable. Church's Thesis is not provable, since it hinges on the intuitive and informal notion of an effective procedure, but it is generally accepted, and it provides a very useful reduction of a fuzzy-but-useful mathematical notion to a crisply defined notion of apparently equal scope and greater power. Analogously, the claim that every mental phenomenon alluded to in folk psychology is *intentional-system-characterizable* would, if true, provide a reduction of the mental as ordinarily understood – a domain whose boundaries are at best fixed by mutual acknowledgment and shared intuition – to a clearly defined domain of entities, whose principles of organization are familiar, relatively formal and systematic, and entirely general.[35]

This reductive claim, like Church's Thesis, cannot be proven, but could be made compelling by piecemeal progress on particular (and particularly difficult) cases – a project I set myself elsewhere (in *Brainstorms*). The final reductive task would be to show not how the terms of intentional system theory are eliminable in favour of physiological terms via sub-personal cognitive psychology, but almost the reverse: to show how a system described in physiological terms could warrant an interpretation as a realized intentional system.

REFERENCES

Block, N. 1978. 'Troubles with functionalism.' *Perception and Cognition: Issues in the Foundations of Psychology*, ed. C. Wade Savage, pp. 261–326. Minnesota Studies in Philosophy of Science, vol. IX. Minneapolis: Minnesota University Press.

Burge, T. 1979. 'Individualism and the mental.' *Midwest Studies in Philosophy*, vol. IV, pp. 73–121.

Campbell, D. 1973. 'Evolutionary epistemology.' *The Philosophy of Karl Popper*, ed. Paul A. Schilpp. La Salle, Illinois: Open Court.

Davidson, D. 1970. 'Mental events.' *Experience and Theory*, ed. L. Foster and J. Swanson, pp. 79–102. Amherst: University of Massachusetts Press.

Dennett, D. C. 1971. 'Intentional systems.' *Journal of Philosophy* 68, 87–106. Reprinted (with other essays on intentional systems) in *Brainstorms*, pp. 3–22.

Dennett, D. C. 1974. 'Why the law of effect will not go away.' *Journal of the Theory of Social Behaviour* 5, 169–187. Reprinted in *Brainstorms*, pp. 71–89.

Dennett, D. C. 1975. 'Brain writing and mind reading.' *Language, Mind and Knowledge*, ed. K. Gunderson. Minnesota Studies in Philosophy of Science, vol. VII. Minneapolis: Minnesota University Press. Reprinted in *Brainstorms*, pp. 39–50.

Dennett, D. C. 1976. 'Conditions of personhood.' *The Identities of Persons*, ed. A. Rorty. Reprinted in *Brainstorms*, pp. 267–85.

Dennett, D. C. 1978. *Brainstorms*. Montgomery, Vermont: Bradford Books; Hassocks, Sussex: Harvester Press.

[35] Ned Block (1978) presents arguments supposed to show how the various possible functionalist theories of mind all slide into the sins of 'chauvinism' (improperly excluding Martians from the class of possible mind-havers) or 'liberalism' (improperly including various contraptions, imagined human puppets, and so forth among the mind-havers). My view embraces the broadest liberalism, gladly paying the price of a few recalcitrant intuitions for the generality gained.

Field, H. 1978. 'Mental representation.' *Erkenntnis* 13, 9–61.
Fodor, J. 1975. *The Language of Thought.* Hassocks, Sussex: Harvester Press; Scranton, Pa.: Crowell.
Grice, H. P. 1957. 'Meaning.' *Philosophical Review* 66, 377–88.
Jeffrey, R. 1970. 'Dracula meets Wolfman: acceptance vs. partial belief.' *Induction, Acceptance and Rational Belief,* ed. Marshall Swain. Dordrecht: Reidel.
Lewis, D. 1974. 'Radical interpretation.' *Synthèse* 23, 331–44.
Nisbett, R. E. and Ross, L. D. 1978. *Human Inference: Strategy and Shortcomings.* Englewood Cliffs, N.J.: Prentice Hall.
Putnam, H. 1975. 'The meaning of "meaning".' *Mind, Language and Reality (Philosophical Papers,* vol. II), pp. 215–71. Cambridge: Cambridge University Press.
Reichenbach, H. 1938. *Experience and Prediction.* Chicago: University of Chicago Press.
Ryle, G. 1949. *The Concept of Mind.* London: Hutchinson.
Schank, R. 1977. 'Sam – a story understander.' Research Report 43, Yale University Dept of Computer Science.
Schank, R. and Abelson, R. 1977. *Scripts, Plans, Goals and Understanding.* Hillside, N.J.: Erlbaum.
Scheffler, I. 1963. *The Anatomy of Inquiry.* New York: Knopf.
Simon, H. 1969. *The Sciences of the Artificial.* Cambridge, Mass.: M.I.T. Press.
Sober, E. (unpublished) 'The descent of Mind.'
Stich, S. 1978. 'Belief and subdoxastic states.' *Philosophy of Science* 45, 499–518.
Tversky, A. and Kahneman, D. 1974. 'Judgement under uncertainty: heuristics and biases.' *Science* 185, 1124–31.
Woodfield, A., ed., forthcoming. *Thought and Object.* Oxford: Oxford University Press.

Time and reality

P. C. W. DAVIES

1. INTRODUCTION

Time is our most elementary experience of the world, so it has a special status in our conception of reality. Eddington once remarked that time enters into our consciousness through a sort of 'back door' and I think it is generally conceded that this so-called psychological time possesses apparent qualities that are absent from the 'outside' world of the laboratory. This additional structure consists of an awareness of a *now* or present moment, and an impression that time passes. Modern physics certainly distinguishes past-facing and future-facing directions of time, i.e. temporal asymmetry in the behaviour of physical systems, but it makes no use of either *the* present (as opposed to a particular moment) or the flow of time.

Newton (1686: 6) described time as passing ('flowing equably' was the expression used), but there is no instrument that can measure its rate of flow (one second per second?) and no physicist that I know who can even make sense of the idea. If our conception of reality is based on our experience of time, therefore, it is seriously at odds with the external world whose reality we are concerned with.

Having said this, there is a fundamental aspect of time on which nature and mind do seem to be in agreement, namely time asymmetry – the distinction, not between *the* past and *the* future (for they presuppose *the* present), but between past and future *directions* of time. Moreover, I am of the opinion that both the nature and origin of this time asymmetry are understood in outline (see Section 4). Lest the distinction between time asymmetry and the flow of time is unclear let me give two analogies. The Earth's surface is asymmetric with respect to north–south by virtue of its rotation and magnetic field. A compass needle reveals the asymmetry and is equipped with an arrow to denote it. The arrow does not imply that we are *moving* north, only that north-direction and south-direction are physically distinct. Which end the head of the arrow points is purely convention. Likewise the so-called 'arrow of time' can point either way and does not imply a motion of, or through, time; only a past–future asymmetry analogous to north–south.

It is also important to realize that we have no physical evidence that time *itself* is asymmetric. Time asymmetry is a property which, as a matter of observational fact, exists in the material world. The *world* is asymmetric *in* time. Without matter, there would be no asymmetry. This is well illustrated by the second of my analogies. Take a movie film of a typical time asymmetric sequence such as an egg falling to the floor and breaking. Cut the film into individual frames, shuffle them and give them to a colleague to sort out. He will easily be able to recognize the correct sequence, being familiar with the temporal asymmetry of the world, and will order the stack correctly. Now the stack itself clearly possesses a structural asymmetry – there is a top and bottom which are structurally distinct (again, *which* way up is purely convention) – which maps in vertical space the temporal asymmetry of the world in the vicinity of the egg. It is not necessary actually to *run* the movie film (make time flow) to discover this asymmetry.

Although modern physics cannot provide an objective basis for a flow of time, in contrast it does introduce new features into time that are beyond our experience and which therefore cause puzzlement or distress. These will now be described.

2. THE BEGINNING AND END OF TIME

Kant notwithstanding, modern physics provides scenarios in which time ceases to exist and, in reverse, comes into existence. This startling possibility is based on Einstein's general theory of relativity that regards time as but one projection of a unified spacetime. Moreover, spacetime is not merely a static arena in which the activities of matter are embedded, but is itself a dynamical entity that can 'change', 'move', and maybe 'shrivel' out of existence. The reason for the dynamical nature of spacetime is that it is used as a description of *gravity* which, being a field, can store and transport energy and is subject to laws of motion, as with the electromagnetic field. A dramatic demonstration of the dynamical nature of spacetime is given by gravity waves, widely believed to be the explanation for the energy loss in the binary pulsar discovered by astronomers in 1974. Gravity waves are literally ripples of spacetime, which can in principle set up sympathetic vibrations in metal or crystal bars.

Thus, in relativity, spacetime is in a sense 'elastic'. Changes in velocity by the observer can cause time to stretch (the famous time dilation effect) as can the proximity of a massive body. These are measurable effects. The expansion of the universe causes space to stretch. In addition to stretching, spacetime can twist (for example around a rotating body), bend and otherwise distort itself in a variety of ways. What we are concerned with here is whether spacetime can tear, shrivel away, or curve so severely that it becomes cusp-like and stops. Any such regions would be considered as boundaries to spacetime. They could be in a spacelike direction, implying temporal extremities, or timelike, implying an abrupt edge to space. Mathematicians call these boundaries *singularities* and many beautiful theorems exist that predict not only their occurrence, but their ubiquity. Let me explain a little more about how these unusual possibilities arise.

There are two scenarios in which spacetime singularities might occur. One is total gravitational collapse. When a star burns out, its enormous weight can no longer be supported by thermal pressure, so it implodes. If it is big enough (say at least three solar masses, which is common) and if no material is lost during the implosion, there is no means whatever by which the star can be supported without violating the theory of relativity. It must continue to shrink without limit. If it is exactly spherical its surface will contract to zero area in a finite (and very brief) time as measured on the collapsing star, so that it rapidly reaches a state of infinite density. Crudely, the gravitational field at the shrinking surface rises like some inverse power of the radius, so the gravity, and hence the curvature of spacetime, also rises without limit as the implosion escalates.

Now, clearly physics cannot survive a profusion of *infinite* quantities, and this includes spacetime. The dynamical equations for spacetime break down, and the concept cannot be continued *through* the singularity. Thus, spacetime comes to an end there. Of course, it could be that the spacetime structure breaks down before infinities occur (of which more later) and some new structure replaces them, enabling physics (of some new sort) to survive gravitational collapse. But if we believe relativity, then either way spacetime as such meets a demise.

It used to be thought that departures from exact symmetry might cause most of the star to miss the exact centre, but in a series of powerful theorems Roger Penrose and Stephen Hawking proved that they are generic features of gravitational collapse, although they may not neces- sarily involve all the imploding matter (Hawking and Ellis 1973). Singularities in spacetime give cause for serious alarm, for they represent a complete breakdown of the causal structure of the universe. Matter and other influences may enter or leave the world through them in a completely unpredictable way. Cosmic anarchy, however, is circum- vented if all singularities are hidden inside black holes. These are regions of spacetime bounded by event horizons which preclude information from the interior reaching the outside world. Matter falling into a black hole can leave spacetime entirely at the singularity within, but influences from the singularity cannot invade the region beyond the black hole. Considerable effort has been devoted to trying to prove the conjecture that 'naked' singularities are forbidden (the so-called cosmic censor hypothesis; Penrose 1969).

The second scenario for singularities is cosmology and the global dynamics of the universe. The present expansion may eventually be arrested, if there is sufficient matter in the universe, and give way to contraction. The rate of shrinkage will accelerate until the universe implodes catastrophically in a gigantic 'big crunch', smashing itself into oblivion at a spacetime singularity. This would represent a future temporal boundary to the whole universe, at which not only all of matter, but space and time as well, cease to exist: no things, no places, no moments. It is important to realize that one is not left with a void, but with nothing at all. The price paid for spacetime to acquire, *via* the theory of relativity, a more concrete (almost tangible) existence is that it might forfeit that existence at a singularity, as might any other dynamical entity. This is the end of the universe and the end of time. The continuity of time stops at a singularity, and the spacetime manifold is no longer differentiable, thus precluding the continuation of any physics based on differential equations.

The time reversal of this obliteration is the big bang, at which the whole universe apparently came into existence about fifteen billion years ago. Once again, I must stress that the big bang is not the appearance or expansion of a blob of matter into a pre-existing void, but the appearance of spacetime itself. Thus, questions about what 'caused' the big bang, or what 'happened before' it, are meaningless. We must face the fact that the big bang singularity is apparently naked and therefore in any case unpredictable. We might expect anything that came out of it to be completely random, a fact which will prove of the greatest significance in explaining the origin of time asymmetry.

A good analogy in the cosmological case is a rubber balloon, the two-dimensional surface of which represents all of three-dimensional space. (Note that this surface is finite, yet unbounded. Our universe could possess a three-dimensional version of this geometry, in which the volume of space is finite, although there is no special centre, or edge, an idea originally suggested by Einstein. If recollapse occurs then space will be of this finite, closed type.) As the balloon is inflated, the surface expands uniformly, like the expanding universe. Note that it does not expand away *from* any particular place (on the surface) or *toward* any place. It merely scales up in size everywhere. Similarly, the universe did not explode *out of* any lump, or centre of creation: the whole of space erupted into expansion everywhere.

Now let the balloon deflate, and imagine it can shrink without limit until it shrivels away to nothing at a point. This is the big crunch. Play it in reverse and you have the big bang. Both are truly the ends of time.

I should like to end this section by remarking that if time did not have a past extremity we would encounter a serious paradox. The universe would have to be infinitely old, and there are problems about why it has not reached a state of enormous entropy (see the section on time asymmetry). A model cosmos has been suggested in which successive cycles of expansion and contraction occur, in which the universe survives its encounter (somehow) with the singularities. There is a fundamental problem with this, however. The total primeval heat content of the universe can be calculated and works out at about the same as that due to the stars. Hence, if the primeval heat is due to the accumulated starlight of all the previous cycles, it ought to be arbitrarily greater than the present starlight, and we should be vaporized. The only resolution would seem to be a universe of at most one previous cycle (i.e. of finite age), or possibly one that contracted from a condition of infinite dispersion.

3. QUANTUM TIME

The realization that spacetime is dynamical rather than immutable raises the question of how it moves. What laws of motion does it obey? Classically Einstein's equations play for spacetime the role of Newton's laws for matter, but just as matter has to be quantized (and also other fields such as the electromagnetic field) so we should expect the gravitational field to be quantized. What does this mean when this field is spacetime itself?

One description of a quantum system is in terms of alternative paths of evolution, all being potential paths, but only one of which will be the actual path when an observation is made. We can construct a space of paths, and regard the quantum state as a general vector in that space which collapses onto a particular vector (i.e. path) when a measurement is made. The crucial quantum feature is, of course, that these alternative paths are not disjoint possibilities, like heads or tails on a flipped coin, but can interfere with each other.

When the paths concern space itself (ultimately the whole universe), we have a space of spaces, called by John Wheeler (1970) *superspace*. We can imagine an infinity of potential worlds – possible spacetimes – only one of which will be found as the actual world on observation. Nevertheless the worlds are not merely distinct alternatives, they *overlap* each other (the interference property). The effect of measurement is to destroy this overlap and isolate one particular member. There are fundamental

67

problems about the measurement process when the quantum system is the entire universe, but I shall not become entangled with that. I wish merely to draw attention to the fact that 'reality' is only meaningful in the context of such a measurement, and that the idea of one single, objectively existing space and time is a fiction.

The superposition of many alternatives and the inherent uncertainty in the dynamical evolution of a mechanical system are fundamental and familiar features of quantum theory and must surely apply to any quantized version of general relativity. On an ultra-small scale (10^{-33} cm, 10^{-43} sec) quantum fluctuations would become so energetic that they disrupt not only the geometry of spacetime, but perhaps its topology. Many physicists believe that the continuous structure of time (and space) break up altogether at those dimensions and that spacetime must be replaced with a more complicated and elaborate structure (something sponge-like?). It has even been suggested that discrete 'chronons' may appear.

In spite of these expected general features of quantum gravity, no proper *mathematical* theory yet exists (Isham 1977, 1979). Indeed, such has been the difficulty of producing one that some physicists believe it is an impossibility, and conjecture instead that spacetime is incompatible with quantum concepts. In this approach, one has a completely new *pre-geometry*, out of which both spacetime and quantum mechanics would grow as suitable limits. One example of an attempt at such a theory is Roger Penrose's (1973) twistor programme. Thus, instead of forcing a marriage between quantum theory and relativity, one attempts to regard the two as sibling children of some pre-spacetime, pre-quantum parent.

4. TIME ASYMMETRY

There are so many ways in which the world is asymmetric in time that a comprehensive explanation might seem hopelessly ambitious. Nearly all the asymmetric processes, however, may be summed up with the phrase that the cosmos is becoming progressively more disorderly. The universe is dying: the disintegration is apparently irreversible.

The subject of order and disorder is best dealt with in the framework of thermodynamics in terms of entropy, but the same principle is apparent in other branches of physics. Other asymmetric phenomena include wave retardation (i.e. ripples travel outwards from a point of disturbance), the collapse of the wave function in quantum mechanics, the expansion of the universe and the decay of the neutral K meson. (The latter two might not be regarded as an increase in disorder.)

The history of time asymmetry is long and convoluted; the con-troversies tedious and recurring (a variant on the celebrated Poincaré (1890) theorem being that misunderstandings about time asymmetry are repeated infinitely often). Many of the technical details are covered in my book (Davies 1974), and all I shall do here is to extend some of the ideas there, and examine the subject from a slightly different perspective.

In spite of the fact that many eminent people (see, for example, Prigogine 1980) think otherwise, I believe there is general agreement that time asymmetry does not reside in microscopic processes of a system *itself*, but in the way the system is formed by its surroundings, i.e. in the initial or boundary conditions. Muddle over the question of the flow of time has led many physicists on a wild goose chase to find a mysterious extra in-gredient, absent in hitherto known physics, to make the world inherently asymmetric as it evolves. This entails putting the asymmetry in by hand.

To make matters concrete, consider a box divided into two halves by a membrane. In the left is gas A, in the right is B. The membrane is removed and the gases diffuse into each other until they are thoroughly mixed up. No further change appears to occur: the system has apparently reached thermodynamic equilibrium and the transition from unmixed to mixed seems irreversible. It is a perfect example of a time-asymmetric change and is clearly a transition from order to disorder.

The issues surrounding this sort of example are well known. The entropy of the gas rises because random collisions break up the initial organization and spread the molecules about chaotically. Boltzmann's (1898) H theorem demonstrates that so long as the collisions do not conspire to miraculously reorder the molecules (i.e. so long as the collisions are chaotic) then, with overwhelming likelihood, the entropy will increase and the gases mix. Nevertheless, the molecular motions themselves are all reversible, and indeed, given long enough, one will expect random fluctuations from equilibrium to occur and temporary spontaneous reorganizations of the gas to result. After an immensity of time the system would revisit its initial state, according to Poincaré's (1890) theorem. To do so the gases must separate themselves and retreat to opposite ends of the box, thereby reversing the apparently irreversible mixing process, and reducing the entropy. Consistency with the H theorem is retained by recalling that the latter is merely a statistical theorem, i.e. very probably, but not certainly, correct.

The situation is similar to that of shuffling cards. Almost any shuffle breaks suit order, and chaotic sequences will prevail for almost all subsequent reshuffles, but eventually a chance arrangement will repro-duce suit order again, given long enough. Of course, there are vastly more small fluctuations which order, say, a short numerical sequence, than the really big, suit-generating ones.

The underlying time symmetry of the individual molecular motions ensures that an isolated box of gas can never be truly time-asymmetric. Given a low-entropy state then we will almost certainly see a subsequent entropy increase, but equally, we will also see entropy decrease – for how otherwise did the gas get into the ordered state in the first place?

In the real world, systems do not become ordered by random fluctuations from equilibrium (the universe isn't old enough for that to happen) but by *outside interference*. We make a box and put the gases into it. Clearly we should look to the formation of the system as (in Reichenbach's (1956) terminology) a branch system from the main environment, for the origin of time asymmetry.

Before tracing this thread of argument let me dispense with a recurring problem. The box of gas couples to the outside world not merely through its formation process, but continually through the walls. The wall atoms pick up influences from the surrounding world and communicate them through to the interior by jiggling about on the inside face and disrupting the motions of the gas molecules that collide with them. It is often alleged (see, for example, Blatt 1959) that this 'cosmic noise' is the origin of true time asymmetry (whatever that means). It is not claimed that the wall jiggles cause the mixing of the gases – that would be absurd – but that there is nothing asymmetric about gas mixing in the absence of cosmic noise. The reason for this claim is that the mixed gases only look more chaotic than the separated state because we are too big and slow-witted to follow the incredibly complex gymnastics of all the individual molecules. It is pointed out that all the information necessary to reconstitute the original (orderly) state is still there, folded up in the gas in an inconspicuous way. That the molecules really do remember their initial state and could find their way back if necessary is evidenced by the fact that a miraculous simultaneous reversal of all velocities would send the system back to its initial configuration. However, if we intercede with the cosmic noise then the information gradually dissipates away through the walls of the box and the system slowly forgets its origin. It is then no longer reversible.

According to these ideas, ordinary mixing (sometimes called coarse-grained mixing) is a purely subjective, anthropomorphic time asymmetry. This macroscopic irreversibility is not true irreversibility at all (which is true, on a long time scale). In contrast, cosmic noise mixes at the microscopic level and is genuinely irreversible (which is also true, unless the entire cosmos reverses).

However, I believe the issue to be philosophical rather than physical. The cosmic noise school seems to me to represent an extreme reductionist standpoint, in which all features of nature are explained with reference to properties of their constituent parts. This view of nature misses much of interest. There surely exist non-linear, and collective, phenomena, that have no meaning at the microscopic level? Are we to suppose that biology is a purely subjective and therefore useless science because we can prove that not a single atom of the reader's body is living, or ever has been living? The quality of *organization* is precisely a collective, macroscopic one, and we have two choices. We can deny that it has any fundamental significance because if we were all atoms we would not notice it, or we can regard our macroscopic, coarse-grained, comprehensive perspective as a positive advantage and take organization seriously. In which case the mixing of gases is a very real event, with or without minute effects of cosmic noise.

Returning to the question of how the gases came to be in their ordered arrangement in the first place, we must obviously ask why the surrounding environment is so ordered, i.e. not in equilibrium. In the case of the Earth, most of the organized activity is driven by the sun, whose disequilibrium is due to the nuclear burning taking place in its core. The reason it is in this unbalanced state is because it is made of hydrogen, whereas the most stable form of matter is iron. This fundamental thermodynamic disequilibrium prevails throughout the universe – hot, dense stars among cold, empty space. The whole system is slowly running down like a clock as the stars burn, turn into iron, and dissipate their energy around the universe. We thus have the curious problem of how the cosmic 'clock' got wound up in the first place. Why isn't the universe already made of iron?

There is a ready answer for these questions. At one time the disequilibrium of the universe was paradoxical because astronomers could not understand why the hot stars hadn't heated up the cold space (Olbers' (1826) paradox). We now know that there hasn't been nearly long enough for this to happen. Modern cosmology pictures a big bang, or creation event, a mere fifteen billion years ago. The origin of the thermodynamic disequilibrium, and hence the arrow of time, resides in the primeval phase.

The famous second law of thermodynamics embodies the principle that order continually gives way to disorder. It follows that as we look back into the past we should see a progressively more ordered cosmos. This is found to be true as far as optical telescopes have access, but when radio telescopes are used an astonishing result is found. At the earliest moments we can explore (about 10^5 years after the big bang) the universe appears chaotic and close to thermodynamic equilibrium.[1] The primeval phase

---

[1] I am referring to the universal cosmic background heat radiation, of thermal equilibrium form, widely held to be the fading glow of the primeval furnace.

shows no evidence of the systematic structure or organization that we observe around us at present. Where has all this organization come from? How has order arisen out of chaos in apparent defiance of the second law of thermodynamics? What is the mechanism whereby information (e.g. the structure of a galaxy or a man) has appeared spontaneously in the universe?

It can be shown that the universe became 'wound up' during the first ten minutes of its existence. Since then it has been steadily unwinding. The key element in this claim concerns the nature of the microscopic processes engaged in during earliest epochs. Before about one second, the temperature was too hot for much material structure to exist. The primeval fluid was expanding explosively fast and rapidly cooling as a consequence, but the density and temperature were so high that the complicated subatomic processes easily kept pace (their relaxation times were much less than the expansion time scale). Therefore, the composition of the cosmological material easily readjusted itself to maintain equilibrium, in spite of the rapidity of global change.

After one second the situation altered. The temperature was low enough for nuclear reactions to begin, but these are many orders of magnitude slower than the expansion rate at that time (which was still fast). Consequently they did not progress very far, and nowhere near far enough to produce any iron. Only helium formed in abundance. Then, after several minutes of half-hearted nucleosynthesis, the temperature fell below that necessary for the light nuclei to overcome their electric repulsion. The nuclear reactions switched off, and most of the fuel remains in a frozen state, as hydrogen, ready to convert to iron wherever the temperature exceeds a few million degrees, e.g. in the centres of stars. Thus the material of the universe is *metastable*. It would like to become iron, but is frozen as hydrogen and helium by the low temperature and the nuclear electric fields.

The winding mechanism is thus traced to the cosmological expansion motion during the primeval phase (the current expansion is too sluggish to have a significant effect). The reason that order has arisen from primeval chaos is that the cosmological material is not really an isolated system, but is embedded in a globally expanding spacetime. By cooling the fluid, the expansion opened the way to starlight through nuclear burning, and thence to all the organized activity of the type that we observe.

Primeval chaos also explains the time asymmetry of electrodynamics, i.e. that electromagnetic waves are retarded rather than advanced. The existence of advanced radiation would require the cooperation of widely separated regions of the universe to emit radiation in such a way as to set up converging, coherent, imploding shells of waves (i.e. the time reverse of the usual outgoing, expanding waves). The probability of this conspiracy occurring in a randomly structured, incoherent primeval explosion are infinitesimally small. Moreover, the absence of primeval conspiracy is beautifully explained by the assumed existence of a singularity at the beginning. Being an inherently unpredictable creature, an initial singularity might be expected to produce only random influences (Hawking 1976).

The most important conclusion of the study of these primeval epochs is that the highly ordered universe we now see did not always exist, but arose *spontaneously* as a result of the expansion. It is now in the process of decaying again.

5. GRAVITY AND BLACK HOLES[2]

Although stellar structure and nuclear physics provide a straightforward account of the rise of order in the universe, they cannot be the whole story. There are disordering processes familiar in daily life that clearly do not have their origin in the sun's nucleosynthesis; for example, the tides which slowly erode the continental shelves. Terrestrial tides are caused by the moon's gravity, and astrophysics provides many other examples of the role of gravity in organizing the behaviour of the cosmos. Gravity shapes the structure of the solar system and the galaxy, and controls the expansion of the universe. We look to gravity to explain the existence of discrete objects, such as stars, planets and galaxies.

Though weak, gravity is a cumulative force which, given enough gravitating matter, can overwhelm all other forces of nature. It acts indiscriminately between all matter and energy in the universe. Being always attractive, it operates universally to pull matter together into ever more compact lumps. In relatively low-mass objects like the Earth, the force of gravity is balanced by atomic forces. In the sun it is resisted by the huge interior pressure generated by the nuclear furnace in the core. In some stars that have expended their nuclear fuel, the pressure support fails and they implode under their own immense weight. For stars with masses comparable to the sun, atomic or nuclear forces may still prevent total collapse, and the imploded star will settle down in a highly shrunken condition as a so-called white dwarf, or as a neutron star. However, if its mass is a few times greater, no known means of support exists, and it is widely supposed that the star will shrink rapidly (in a few microseconds, star-time) and without limit.

Whatever happens to the star, its surface gravity becomes so intense

[2] This section is adapted from an article in *The Great Ideas Today*, 1979, pp. 43–55. It appears here by permission of Encyclopaedia Britannica, Inc.

that it seriously distorts the structure of space and time in its vicinity, trapping all light and other objects inside a surface known as an event horizon. The black, empty region inside the horizon is called a *black hole*. Many astronomers believe that most heavy stars will end their days inside a black hole.

Black holes represent the ultimate triumph of gravity. Objects like the sun and stars are really only a temporary, metastable interlude between a distended cloud of gas and totally imploded matter. This universal tendency of gravitating systems always to shrink is reminiscent of the second law of thermodynamics, and in recent years some spectacular discoveries by mathematicians and theoretical physicists about black holes have confirmed the close connection between the irreversible ubiquitous rise of entropy and the irreversible onslaught of gravitational collapse. (For a review, see Davies 1978.)

One easy way of understanding why a black hole is a highly disordered object is to appeal to the concept of entropy as the negative of information (Shannon and Weaver 1949). When a black hole swallows matter, it becomes imprisoned inside the event horizon. From the outside world all information about the sacrificed matter is lost, except for its total mass (the black hole grows a little larger), and any electric charge or rotation (which persists on the hole). Thus, the black hole destroys information more efficiently than anything else known. For example, it is not possible to distinguish whether a black hole was formed from an imploded star, or antimatter, or light, or green cheese. The information content of the matter (e.g. its cheesiness) gets wiped out when it falls across the horizon. It follows that there are a vast number of ways of making a black hole of given mass, charge and rotation. Thus, just as a box of gas is maximally disordered when its state can be achieved in the greatest variety of ways, so too a black hole is the maximally disordered *gravitating* system.

We may enquire how all the objects we see in the universe have avoided becoming black holes (observations suggest that black holes cannot form the overwhelming majority of astronomical objects). Just as matter in the absence of gravity would like to be iron, so gravitating matter would like to be a black hole. Is it possible to explain why the big bang did not cough out only black holes in the same way that we explained why it did not present us with an iron universe?

The problem here is that, whereas when we consider earlier and earlier moments of the primeval fireball we can be more and more sure that iron did not exist (the temperature becomes progressively higher at earlier moments), we cannot say the same about black holes. Indeed, crudely speaking, the more compressed material at earlier epochs was more, rather than less, likely to form these objects. Thus, the importance of black hole formation rises along with our ignorance of the physics as we probe back towards the totally mysterious circumstances near the first moment. This makes any attempted explanation of how so much matter escaped falling into black holes extremely conjectural. We know almost nothing about ultra-hot, ultra-crushed matter. It is conceivable that it was so stiff that even when compressed very close to black hole densities, most of the material resisted.

An amusing calculation is to estimate just how improbable it is that an overwhelming black hole population is absent by chance (Penrose 1979). This may be done by comparing the entropy of a world full of black holes with that of the entropy we actually observe. Most of the observed entropy is contained in the primeval heat radiation. If the entire observable universe were a single gigantic black hole, its entropy would exceed that of the cosmic heat by the colossal factor of 10^{30}. Moreover, the relationship between entropy and probability is an exponential one, so this number translates into the absurdly minute odds of one chance in about $10^{10^{30}}$ (that is one followed by 10^{30} zeros!) against a chance origin for the present hole-less arrangement. Either the universe is a 'miracle', or some as yet unknown mechanism stiffens the primeval matter.

If this latter conjecture is correct, it may explain another of the great order–disorder mysteries of cosmology. It has long been wondered why the expansion of the universe is so *uniform*. As far as we can tell, whichever way we look into space, the rate of expansion is the same. This type of uniform expansion corresponds to a universal *dilation* of space. That is, the expansion is not haphazard or directional, but behaves just as though the *scale* of cosmic distance were gradually stretching everywhere and in all directions at the same rate. Why, out of all the ways in which the universe could have expanded, did it choose such an orderly pattern of motion? In the primeval phase had the expansion energy been shared randomly among all the available channels of expansion, we would have ended up with a highly turbulent, chaotic expansion. Instead, nearly all the energy has been concentrated into a single, dilatory mode.

There are two objections to the use of this type of semi-quantitative statistical reasoning. The first is that there is no compelling choice of probability measure to be imposed on the space of initial cosmological data. The second is that Hawking's (1975) formula for black hole entropy makes no direct use of statistical concepts, being a purely geometrical expression involving event horizon area. It seems that further progress in estimating the 'unlikelihood' of the observed gravitational arrangement must await a full theory of the thermodynamics of self-gravitating systems, which itself will probably require a decent theory of quantum gravity.

In recent years some attempt has been made to explain the absence of turbulent disorder in the cosmic motion as due to the dissipation of the turbulent energy as heat in the primeval fireball. One problem is that the dissipation mechanisms investigated seem to be progressively more efficient at earlier epochs. But it can be shown that the heat produced for a given amount of turbulence rises without limit the earlier it is delivered (Barrow and Matzner 1977). Thus we would expect even a minute amount of turbulence to generate enormous heat. Moreover, we know that the universe is actually very cold (about three degrees) which makes it seem unlikely that there was ever very much turbulence at all. Once again we are faced with the mystery of why the real universe chose such a remarkably orderly path of behaviour from among overwhelmingly many neighbouring paths that lead to chaos. If the primeval material possessed strange stiffness properties preventing local conglomerations into black holes, this might also account for how it avoided disintegrating into chaotic expansion.

6. REALITY, IRREVERSIBILITY AND QUANTUM MEASUREMENT

So far I have said nothing about the other important example of time asymmetry – the so-called collapse of the wave function which occurs during the measurement process of a quantum mechanical system. The subject is frequently muddled by claiming that a mystery surrounds how the system changes from a superposition of many alternative potential results to one actually measured result. This is not the real issue. Even in classical physics we must accept the existence of *mixed states* in general because we have only limited information in practice about the state of a system. If we toss a coin and do not read the outcome, it is in a mixed heads–tails state. That means *either* heads, or *tails*, each with a 50% probability, but not *both*. When we actually look to see, in a sense the state 'collapses' from heads/tails to either heads or tails as the case may be. This process is, of course, irreversible (we cannot 'unread' the result) but the asymmetry is of a familiar classical kind to do with the accumulation of information and the entropic rearrangement of the brain.

The quantum collapse is altogether more profound, for the quantum state is not a collection of either–or alternatives, but a *superposition* in which these alternatives *overlap* each other (see, for example, d'Espagnat 1971, or Davies 1980). The different possibilities combine *coherently*, i.e. they have interference terms. Thus, an uncertain quantum state differs fundamentally from an undetermined classical state. The effect of coupling the quantum system to some macroscopic apparatus results in the destruction of the interference terms. The state changes from a coherent superposition to an incoherent mixture, similar to heads/tails. This process is irreversible, whether or not the experimenter chooses to reduce the result further by looking at his measuring apparatus to ascertain which particular outcome had been recorded. It amounts to chopping all the overlapping potential worlds into disconnected alternatives.

Quantum measurement is thus a two-stage process. In the first stage, the wave function (state vector) changes from a superposition to a mixture, and in the second step the mixture further collapses to a particular value if the apparatus is read. The second step is easily explained in terms of ordinary thermodynamics, the first is not and is genuinely a *microscopic time asymmetry*.

I refer the reader to my books (Davies 1974, 1980) for a detailed discussion of these issues. Here I wish merely to remark that any definition of reality must be based on the *mixed* state and not the superposition. I believe this to be the conventional (Copenhagen) picture as expounded by Niels Bohr (1934). Reality only comes into existence upon the collapse of the wave function from a superposition to a mixture. As this step is the *only* place where quantum systems are microscopically time-asymmetric (i.e. asymmetric even in the absence of coarse-graining – I ignore here questions of K mesons) the origin of the time asymmetry goes hand in hand with the origin of reality. Thus the nature of time is strongly connected at the quantum level with the nature of reality.

This raises intriguing questions about the reality of the entire universe. The well-known von Neumann chain (1932, ch. VI), in which the wave function for the total system (microscopic + measuring apparatus) itself will only collapse into 'reality' when a subsequent external measurement is made, eventually encompasses the whole cosmos. There is no 'outside' apparatus to measure it. Beside the philosophical challenge, this question raises deep practical issues about quantum cosmology – the application of quantum mechanics to the global dynamics of the universe, which is vital to any proper investigation of whether it will survive its encounter with singularities.

REFERENCES

Barrow, J. D. and Matzner, R. 1977. *Monthly Notices of the Royal Astronomical Society* 181, 719.

Blatt, J. M. 1959. *Progress of Theoretical Physics* 22, 745.

Bohr, N. 1934. *Atomic Theory and the Description of Nature*. Cambridge: Cambridge University Press.

Boltzmann, L. 1898. *Lectures on Gas Theory 1896–1898*, transl. S. G. Brush. Berkeley: University of California Press, 1964.

Davies, P. C. W. 1974. *The Physics of Time Asymmetry*. London: Surrey University Press.

Davies, P. C. W. 1978. *Reports on Progress in Physics* 41, 1313.
Davies, P. C. W. 1980. *Other Worlds*. London: J. M. Dent.
D'Espagnat, B. 1971. *Conceptual Foundations of Quantum Mechanics*. Menlo Park.
Hawking, S. W. 1975. *Communications in Mathematical Physics* 43, 199ff.
Hawking, S. W. 1976. *Physical Review* D14, 2460.
Hawking, S. W. and Ellis, G. F. R. 1973. *The Large Scale Structure of Spacetime*. Cambridge: Cambridge University Press.
Isham, C. J. 1977. *Annals of the New York Academy of Sciences* 302, 114.
Isham, C. J. 1979. *Proceedings of the Royal Society* A 368, 33.
Newton, I. 1686. *Principia*, ed. F. Cajori. Berkeley: University of California Press, 1934.
Olbers, H. W. M. 1826. *Edinburgh New Philosophical Journal*, April–October, 141.
Penrose, R. 1969. *Revista Nuovo Cimento*, 1, 252.
Penrose, R. and MacCallum, M. A. H. 1973. 'Twistor theory: an approach to the quantization of fields and spacetimes.' *Physics Reports* 6C, no. 4, 242–315.
Penrose, R. 1979. In *General Relativity: an Einstein Centenary Survey*, ed. S. W. Hawking and W. Israel. Cambridge: Cambridge University Press.
Poincaré, H. 1890. *Acta Mathematica* 13, 1ff.
Prigogine, I. 1980. *From Being to Becoming: Time and Complexity in Physical Sciences*. Freeman.
Reichenbach, H. 1956. *The Direction of Time*. Berkeley: University of California Press.
Shannon, C. E. and Weaver, W. 1949. *A Mathematical Theory of Communication*. University of Illinois Press.
von Neumann, J. 1932. *Mathematical Foundations of Quantum Mechanics*, transl. R. T. Beyer. Princeton, N. J.: Princeton University Press.
Wheeler, J. A. 1970. 'Superspace.' *Analytic Methods in Mathematical Physics*, ed. R. P. Gilbert and R. Newton. New York: Gordon and Breach.

McTaggart, fixity and coming true[1]

D. H. MELLOR

I. INTRODUCTION

Some events are past, some present and some, I expect, are still to come. These are at once the most obvious, the most basic and the most disputed facts about time. I am one of those who dispute them. I maintain with McTaggart (1908; 1927: ch. 33) that in reality nothing is either past, present or future. Since, however, I part from him by thinking that reality need not be tensed to be temporal, I am not led, as he is, to deny the reality of time itself. Indeed I believe that, paradoxically, time needs to be both real and tenseless to explain how and why people come to think of events as being past, present and future.

These propositions are, I fear, still contentious, so they will have to be defended in what follows. But my main object is not merely to promote and sugar McTaggart's pill. I want also to prescribe it: specifically, for R. C. Jeffrey's 'conceit that the world grows by accretion of facts'; or, in other words, that only when an event happens does the proposition saying so 'come true' (1980: 253). It will also serve to purge J. L. Mackie's closely related conceit that events become 'fixed and settled and unalterable' (1974: 178) as soon as their 'preceding sufficient causes . . . have occurred' (181). These are serious conceits, though not new ones: McTaggart himself (1927: §337) appeals to the second while disposing of Broad's (1923: ch. II) version of the first. But as they have been newly reconceived, so they need renewed purgation. They are, I shall argue, only trivially true if time is tenseless. And rather than tax my distinguished colleagues with triviality, I prefer to conclude that they are wrong.

[1] This paper developed out of classes given at Stanford University in the Fall of 1978, during a visit made possible by the grant of a Radcliffe Fellowship and a British Academy Overseas Visiting Fellowship, for which I owe thanks to the Radcliffe Trust and the British Academy. I am indebted for helpful comment and criticism to several Stanford students, to Professors John Perry, Nancy Cartwright and David Lewis, and to those taking part in the March 1979 meeting of the Thyssen U.K. Philosophy Group, at which the original version of it was discussed. In rewriting it for this volume I have been further assisted by the replies of Professor Jeffrey and Mr Mackie to my critique, and also by detailed comments from Jeremy Butterfield.

2. TIME WITHOUT TENSE

First, however, we must get rid of tense, and I will not pretend that this is easy. Consider for example the fundamental relation '. . . is earlier than . . .' (or its converse, 'later than'). What makes this relation temporal? One persuasive answer is: one event, e, being earlier than another, e', implies such tensed facts as that sometime e' is present and e past but never *vice versa*. What makes the 'earlier' relation temporal, in other words, is that it determines the order in which the events it relates become successively present and then past. But if there are in reality no such tensed facts as events being present or past, something else must make 'earlier' temporal – and it is no easy task to find something else that will do the job. As McTaggart saw, it is not enough for a tenseless relation between events merely to reproduce the order in which they appear to become present. If, for example, everything in the universe was always at the same temperature at the same time, but always cooling, the 'cooler' relation would do that: but that would not make 'cooler' a temporal relation.

Advocates of tenseless time have, I admit, mostly shirked the task, e.g. of saying what is temporal about the non-spatial dimension of their four-dimensional Minkowski manifolds. Their 'block' universes have no more real time in them than McTaggart's does – the difference being that McTaggart sees this and they, by and large, do not. I too will shirk the task here, but I do acknowledge it, since I am not willing to give up real time, and I undertake to tackle it elsewhere. All I can say here is that the materials I will use are the direct perception of one event being later than another, which occurs whenever we see something move or change in some other definite way, and the role causation plays in that perception.

There is another task, however, which I must attempt here: namely, to give a tenseless account of change. Time is essentially the dimension of change, and any theory of time has to account for that fact. Now McTaggart thought that change needed tense, since he thought change to be impossible without events moving from the future *via* the present to the past, a movement I shall call 'McTaggart change'. Without real tense, of course, McTaggart change does not exist, so a tenseless account of change must find a way of doing without it. My account derives from Russell (1903: §442): 'Change is the difference, in respect of truth or falsehood, between a proposition concerning an entity and the time T, and a proposition concerning the same entity and the time T', provided that these propositions differ only by the fact that T occurs in the one where T' occurs in the other.' This is what Geach has called 'Cambridge change' and, as he says, actual change is only one species of it (1979: 90–2). To adapt an example of McTaggart's (1927: §309), 'the fall of a sand-castle on the English coast' effects a Cambridge change in the Great Pyramid, by changing a relation in which it stands to the sand; but clearly the Pyramid itself does not actually change as the sand does. The difference between actual and what Geach calls 'merely' Cambridge change is causal: actual changes are events, with spatiotemporally contiguous effects, and merely Cambridge changes are not.

I follow Davidson (1969) in taking events, including changes, to be individuated by their causes and effects. But not all events are changes; nor do events themselves change. Change occurs in things, i.e. individual substances, in one standard sense of that term. (The difference between things and events I take to be that whereas events, if extended in time, have temporal parts, things do not. People are things in this sense, and so are common objects such as tables, chairs – and McTaggart's (1927: §313) poker. For a longer list, and some reasons why the thing/event distinction matters, see my 1982: §6.) A thing may have a non-temporal property at one date incompatible with those it has at earlier or later dates; and when such a fact constitutes an event, with effects spatiotemporally contiguous to the thing, the thing has undergone an actual change between these dates. We may indeed use this as a criterion for distinguishing real from merely apparent properties of things, thus ruling out such spurious properties as being forty, famous, the tallest man in the room and 'grue' (Goodman 1965: ch. III). Real properties of things and people, loss or gain of which is actual change in them, rather than the merely Cambridge variety, include temperatures, masses, colours, shapes – and both physical dispositions such as solubility (Mellor 1974: §I–II), and mental states such as particular beliefs and desires (Mellor 1978: §II).

Now suppose some thing, a, has a pair, G and G^*, of such incompatible real properties (e.g. temperatures) during two separate stretches of time t and t^*: i.e.

$$a \text{ is } G \text{ during } t \qquad (1)$$

and

$$a \text{ is } G^* \text{ during } t^* \qquad (2)$$

If a were an event, it would have different temporal parts containing a-during-t and a-during-t^*, and the supposed change would reduce to these different parts having different properties:

$$G(a\text{-during-}t) \qquad (3)$$

and

$$G^*(a\text{-during-}t^*) \qquad (4)$$

81

But that different entities differ in their properties does not amount to change, even if one is earlier than the other and both are parts of something else. (3) and (4) would no more constitute a case of change than would *a*'s spatial parts differing in their properties – e.g. McTaggart's poker being hot at one end and cool at the other.

I take change to require a difference between the state of a *whole* thing at two different times. That is, real changeable non-temporal properties of a thing are in fact relations it has to the various times and stretches of times at which it exists. I.e. (1) and (2) should be read as

$$G(a, t) \tag{5}$$

and

$$G^*(a, t^*) \tag{6}$$

Treating temperatures, colours, shapes etc. as relations between things and times may seem odd, but it is only a way of making two indubitable points about facts like (1):

(a) Both the contexts

'. . . is G during *t*'

and

'*a* is G during . . .'

are transparent, i.e. (1) remains true however *a* and *t* are referred to.

(b) For (1) to be true, both *a* and *t* must exist. (This need not of course imply a Newtonian conception of absolute time: it does not follow that time could exist without events – times may still need specifying by events, such as Christ's birth, and their temporal relations, such as the earth's period of rotation on its axis and about the sun.)

I should emphasise at once that (5) and (6) in no way beg the question against tenses. Nothing prevents *t* and *t** taking tensed values like 'yesterday' and 'tomorrow' as well as tenseless ones like '9 January' and '10 January'. Nor do (5) and (6) conflict with the use of sentential operators which Prior's work has made usual in tense-logic; i.e., in this case,

$$\text{During } t, \, Ga \tag{7}$$

and

$$\text{During } t^*, \, G^*a \tag{8}$$

On the contrary, a relational reading of tensed facts is standardly used to supply 'semantics' for these operators (McArthur 1976: ch. 1.3). In other words, even tense-logicians take (7) and (8), with appropriately tensed *t* and t^*, to be made true by the corresponding relational facts as stated in (5) and (6).

However, as an advocate of tenseless time, I will restrict t and t^* to tenseless values. Change, I maintain, consists in a thing's having a real non-temporal property at one date which it lacks at others, i.e. respectively having and lacking, to those dates, the corresponding real non-temporal relation.

McTaggart would not agree; but not because he disputes – he does not draw – my distinctions between things and events and between actual and merely Cambridge change. For McTaggart, (1) and (2), however construed, would not constitute change because they are themselves unchanging facts about a. His poker being 'hot on a particular Monday' and cool thereafter (1927: §315) is no change in it, he says, since it always was and always will be a fact that it is hot that Monday and cool thereafter. And as McTaggart says, neither this 'nor any other fact about the poker change[s], unless its presentness, pastness, or futurity change'. McTaggart change, in other words, is the only kind of change tenseless facts are capable of. But why, in order for a change to be a fact, must that fact also change? I see no reason to believe it must, nor hence any good argument from real change to McTaggart change and hence real tense. We can quite well deny both, and still insist that McTaggart's poker changes as it cools.

3. TENSES AND DATES

We can, I believe, account for time and change without real tense: but why should we try to? Because real tense implies McTaggart change, and that, as he showed, is a myth – the 'myth of passage' as it has been called (Williams 1951). But it is a very powerful myth, and undoubtedly expresses something real and important about time. As the persistent rejection of McTaggart's own sound and simple disproof of it shows, its grip will not be broken until something better is put in its place. In what follows, therefore, I shall put up a tenseless surrogate for it; to which end, I must first lay down more precisely the specification the surrogate has to satisfy.

The myth of time passing, i.e. of McTaggart change, combines two ways of locating events in time: by their dates, and by their temporal distance, past or future, from the present. These two ways locate events in two series of temporal positions which McTaggart called the 'B series' and the 'A series' respectively. McTaggart change consists in the relative motion of these two series. Events of given date become less future or more past, as the present time moves from earlier to later dates.

(There may in fact be several A and B series. In both, events get the same location just in case they are simultaneous; and relativity theory may make the simultaneity of distant events depend, within causal limits, on an arbitrary choice of a so-called 'reference frame', to settle what is to count as being at rest. Physical fact may fail to settle that question: so different but equally good reference frames may make quite different celestial events simultaneous with the terrestrial events of 1 January 1984, for example, thus filling that B series position quite differently. But the same goes for the A series: whatever celestial events get that terrestrial date will *ipso facto* then count as temporally present. So there is, as McTaggart conjectured (1927: §323), a distinct A series corresponding to each distinct B series. For present purposes, however, I can afford to ignore these relativistic complications, since I am concerned only with the apparent relative movement of corresponding A and B series. In referring to 'the' A and B series, then, I shall henceforth mean any relativistically acceptable B series, and the A series corresponding to it.)

Positions in the B series I shall call 'dates', stretching that term to cover locations of all sizes from nanoseconds to millennia. Thus B.C. is a date, and so is the first p.m. second of 1 January 1984. (Events have any date that includes all their temporal parts, just as things have any spatial location that includes all their spatial parts. Thus, the end of World War II has, *inter alia*, the dates A.D., the twentieth century and 1945, just as London has the locations Earth, Europe and England. When I refer to 'the' date of an event, I mean the shortest date that includes all its temporal parts.) Dates may be regarded as intervals of B series instants, such as noon precisely on 1 January 1984, ordered by the 'earlier' relation. I do not of course mean by this that instants exist: if there are any such things, they will be spatiotemporal entities – spacetime points – not purely temporal ones. Instants are no more than convenient theoretical devices for generating indefinitely divisible systems for dating events.

Positions in the A series I shall reluctantly follow custom and call 'tenses', though they are mostly marked, not by verbal inflection but by adverbs and phrases such as 'today', 'ten days hence' and 'last year'; and given these, verbal tenses are redundant – 'last year' already implies the past tense, as 'today' implies the present. Tense in the sense of A series position must therefore be sharply distinguished from verbal tense, which is merely one very crude way of marking it; and the former, not the latter, is what I shall mean by 'tense' hereafter unless I explicitly say otherwise.

Tenses, like dates in the B series, may be regarded as intervals of instants, and these are likewise ordered by the 'earlier' relation. McTaggart (1927: §305) characterises the B series as ordered by 'earlier', as opposed to the A series, which is ordered by degrees of pastness or futurity; but this is a false contrast. 'Earlier' orders both series. Ten days ago, an *A* series position, is earlier than today in just the same sense in which 1 January is ten days earlier than 11 January. In fact, the *A* and *B* series have exactly the same temporal structure. They use the very same 'earlier' relation to order the very same collections of simultaneous events. Fix which *B* series instant is the *A* series' present instant, and either series is immediately definable in terms of the other. To every *B* series instant there then corresponds the *A* series instant which is that much earlier or later than the present instant; and hence to every date, i.e. interval of *B* series instants, there corresponds a tense, and *vice versa*. Thus, when it is now noon on 1 January 1984, 10 a.m. is two hours past, 11 January is ten days hence, and the next century is the twenty-first.

Seeing that the *A* and *B* series are so similar, and so simply interdefinable, what is the difference between them? The difference is that whereas an event's dates are fixed, its tenses are not. By this I mean that its tenses vary with time (this of course being just what McTaggart change is), and its dates do not. Suppose for example that it is now May 1984 and the Queen is 58. That is, she was born 58 years ago; in other words, that event has the tense: 58 years past. The tense of this event obviously varies with time: in 1974, the Queen was only 48 years old; in 1994, she will be 68. Note that the event's tense varies just the same if the time itself is reckoned by tense rather than by date: thus, ten years ago, the Queen's birth was 48 years past; ten years hence, it will be 68 years past. These facts, of course, follow from each other, the general study of such temporal entailments being the business of so-called 'tense-logic'. The reason there is no comparable 'date-logic' is simply that an event's dates, unlike its tenses, do not vary with time, whether the time be reckoned in tenses or dates. The fact now, in May 1984, is that the Queen was born on 21 April 1926; and that always was and always will be the date of her birth. (Some indeed think that before 1926, when the Queen's birth was future, it did not yet exist, and so had no date at all. But no one thinks it ever had, or ever will have, any date other than 21 April 1926.) Date-logic, then, is not studied, because it is too simple. Temporal operators, be they dated or tensed, and however they are iterated, have no effect at all on the classical truth value (if any) of '*e* occurs at *T*'.

Dates, unlike tenses, are outright, temporally unqualified properties of events. That is the essential characteristic of the *B*, as opposed to the *A*, series – and why, provided tenseless sense can be made of 'earlier', it is the fundamental series. The *B* series is definable as the definite temporal structure of all the world's events (on a relational view of time), or of all instants (on an absolute view). The *A* series is neither: it has to be defined in terms of the *B* series plus a present instant. And the present instant has to move: there has to be McTaggart change, or the *A* series would be identical with the *B* series. Past, present and future, therefore, as aspects of reality, stand or fall with McTaggart change. They fall – as we shall see in the course of constructing something tenseless to put in their place. But first let us look at the reasons that support them.

4. TENSED TRUTH, TENSELESS FACT

There are two chief reasons for believing in real tense, and in particular in a real present. One is experiential, the other linguistic. The former is what many take to be an irreducible experience of events being present as they happen to us (or, in the case of actions, as we perform them); in other words, its sheer presentness seems to be an undeniable part of our every experience. A credible surrogate is needed for this. To produce it, however, I must first dispose of the latter, linguistic reason for believing in real tense: namely, that our judgments about the tenses of events are generally either objectively true or objectively false, and real tenses are needed to make them so. In May 1984, for example, it is objectively true to think or say that the Queen is 58. What makes that true seems to be that she *is* then 58, i.e. that at that date her birth really does have the tense: 58 years past. But if reality has no tense, there is no such fact, and we must give this indisputably objective judgment alternative tenseless truth conditions. And once that has been done, explaining away the apparent presentness of our experience will turn out to pose no great problem.

The truth conditions I need are really quite obvious, and also quite indisputable. Even if events have tenses, it turns out that these have nothing to do with making what I shall call 'tensed judgments' about them true or false. The truth value of a tensed judgment is determined entirely by how much earlier (or later) it is than the event it is about. A judgment that the Queen is N years old, for example, is objectively true just in case its date is between N and $N + 1$ years later than that of her birth. It is quite immaterial whether the Queen's birth, or the tensed judgment about it, is past, present or future.

The truth conditions of all tensed judgments are fixed in reality by dates. A present tense judgment is true if, and only if, it differs no more in date from the event it is about than the span of tense it ascribes to that event. E.g. '*e* occurs today' is true just in case it is said or thought on the same day as *e;* '*e* occurs this week' just in case it is said or thought the same week. Past and future tense judgments are true if and only if they have dates as much later or earlier respectively than the events they are about as the tenses they ascribe to them are than the present. Tensed judgments can of course be more complex than the simple ascription of an *A* series position to an event. There are, for example, the judgments commonly expressed in English by verbal tenses such as the future perfect. But the truth conditions of these too are fixed by how much earlier or later their dates are than those of events they are about and other dates definable from these. 'Next year the Queen will have reigned 33 years', for instance, is true just in case the Queen is still Queen the year after that judgment is made, and that year is 33 years later than her accession. And similarly for tensed judgments of any complexity. The real usefulness, indeed, of the standard 'semantics' of tense-logic referred to in Section 2, is that it shows how to derive any tensed judgment's truth conditions from its date in this sort of way.

Dates are not only sufficient to fix the truth conditions of tensed judgments, they are also necessary. Suppose a tensed judgment, e.g. that the Queen is 58, had no date – being, perhaps, one of God's judgments if, as some have said, He is 'outside time'. What could make it true? Not that the Queen really is 58 when the judgment is made; for, given that the Queen was born in 1926, that gives the judgment a date, namely 1984. Without a date, in short, a tensed judgment has no definite truth conditions; and with one, its truth conditions contain no tenses. These facts seem to me to make the idea of real tense not merely redundant, but incredible. Try to suppose that there really is in 1984 such a tensed fact as that the Queen is 58. This supposed fact turns out to be no part of what makes the corresponding judgment true: what does that job is simply the date of the Queen's birth being 58 years earlier. Now a fact which has nothing to do with making any tensed judgment true is surely no tensed fact. But these supposed facts are by definition tensed. Yet in reality no such supposedly tensed facts make any tensed judgment true. So I conclude that in reality there are no such facts: there is no real *A* series, and therefore no McTaggart change.

5. EXPERIENCE AND INDEXICALS

But what then of our experience of tense and of McTaggart change? Tenseless truth conditions seem not to dispose of that. Consider Prior's famous example: 'Thank goodness that's over!', said after a painful experience (Prior 1959). 'That's over' is indeed true if and only if said or thought later than whatever experience the 'that' refers to. But why thank goodness for such a tenseless fact, which could be recognised as such at any time, before or during, as well as after, the pain in question: surely the thanks are given in sheer relief for the pain's becoming past and thereby ceasing to be present?

Not necessarily. 'Thank goodness' certainly expresses relief, and is thus appropriately said or thought just when relief is appropriately felt. But when is that? Prior says it is when a pain is past, as opposed to present or future; whereas I say it is just after the pain, as opposed to during or before it. I cannot see that Prior's tensed account of when relief is appropriate is any better than my tenseless one. And mine does make sense of the whole remark: since 'thank goodness', said of a pain, is appropriate just when 'that's over', said of it, is true, it is always right to say both (or neither) at the same time.

This account of Prior's case gives the clue to a tenseless analysis of the apparent presentness of experience. Like his case, it involves self-consciousness; only here one is making tensed judgments of experience as it occurs, rather than afterwards. Now simultaneity with its subject matter is the defining truth condition of a present tense judgment, as opposed to a past or future tense one; so if I am thinking of my actions or experiences as happening *while* I am thinking of them, I am *ipso facto* thinking of them as being present. And that, I suggest, is all there is to the much vaunted presentness of our experience. Experiences in themselves, like events of every other kind, are neither past, present nor future. It is only our simultaneous consciousness of them, as being simultaneous, which necessarily both has, and satisfies, the tenseless truth conditions of present tense judgments.

Our being trapped forever in the present is not a profound metaphysical constraint on our temporal location: it is a trivial consequence of the essential indexicality of tensed judgment. It is like everyone being condemned to be himself and, wherever he is, to being – as he sees it – here. The judgments 'I am X' and 'Here is Y', made respectively by person X and at place Y, are as objectively and inevitably true for all X and Y as 'It is now T', made at time T, is for all T: but not because X and Y have respectively such real properties as 'being me' and 'being here'. Obviously there are no such personal and spatial equivalents of our supposed tensed facts; and if there were, they would, like tensed facts, be no part of what makes the corresponding judgments true. 'I am X' is true if and only if X judges it; 'Here is Y' is true if and only if it is judged at Y. So anyone who judges, of the place that he is at, that it is here, is bound to be right, wherever he is; and similarly, *mutatis mutandis,* for judgments of one's own first person identity. That is all the inescapability of being oneself and being here amounts to: and so it is with the inescapability of the present.

I conclude that neither our experience of time nor the objective truth of tensed judgments requires, or indeed admits of, real tense. Tensed judgments are simply a kind of indexical judgment, with tenseless truth conditions. But this does not mean either that tensed judgments themselves are really tenseless, or that we could do without them. Tense may not be an aspect of the world; but, as Perry (1979) has shown, it is, like personal and spatial indexicality, an irreducible and indispensable aspect of our thought.

That a tensed judgment is not equivalent to any tenseless one is easily seen. If it were, it would be equivalent to the tenseless judgment that its own truth conditions obtain. For example, a particular judgment J, a 'token' of the 'type' 'It is now T', is true if and only if it is made at T. Let J' be the tenseless judgment that this is so, *i.e.* 'J is made at T'. J is true if and only if J' is. But they are not the same judgment. In particular, if J' is true at all, it is true whenever it is made, whereas J is only true at T.

In other words, as upholders of tense have rightly insisted, tensed truths cannot be translated into tenseless ones. Neither the sentence type 'It is now T', nor Prior's 'Thank goodness that's over', nor any other tensed sentence type, means the same as any tenseless sentence. That is because tensed sentence types are indexical: it is part of their meaning that the truth conditions of their tokens vary with time, which is not true of tokens of tenseless types. But there is no tense in the truth conditions themselves; just as the truth conditions of tokens of 'Here is Y' are (literally!) neither here nor there, despite its being different from any non-indexical spatial judgment.

Not only is indexical judgment untranslatable, it is also indispensable. To suppose that we could make do with a tenseless language is as much a mirage as is real tense itself. Suppose I want to do something at T. Some change in my state of mind is needed to prompt me to act at T rather than some other time. The change of course is my coming to judge 'It is now T', where before I judged, 'It is not yet T.' And for this kind of change of tensed belief there is no tenseless substitute. Because the truth value of tenseless beliefs does not change with time, mere lapse of time gives no cause to change them. But it does give us cause to change our tensed beliefs if we are to keep them true, which it is the object of all our belief to be. And these changes, especially changes of belief from the future to the present tense, are the immediate and indispensable causes of our actions. Whether they cause us to act in time is of course another matter: our mental clocks are as fallible as any others. But without them, i.e. without making tensed judgments, we should have no cause to act at all.

This is my surrogate for the myth of passage: the tensed judgments we need to have, and therefore continually to change, in order to be capable of timely action. This is the truth behind the myth. The error is to misread the tense of these judgments as part of their non-indexical content, and hence to see it as an extra, ever-changing aspect of the objective world.

Having exposed the error, we may hope at last to break the myth, and begin to repair the havoc it has wreaked in the philosophy of time.

6. FIXITY AND COMING TRUE

Tense has not wreaked all its havoc under its own name. Jeffrey's conceit of propositions about events 'coming true' as the events happen, is stated explicitly in tenseless terms; and Mackie's, of events acquiring 'fixity', easily can be. Nonetheless, these specious happenings are nothing if not kinds of McTaggart change. Without real tense they are trivial; and with it, impossible, as I will now attempt to show.

Jeffrey gives events no tenses, only dates; but says that before the date of an event its happening is no fact. In other words, the corresponding tenseless proposition is not then true; though it may be 'ineluctable', if its 'final truth' is determined by the facts to date. As time goes on, therefore, propositions come true, and the number of facts increases: 'the world grows by accretion of facts'. What is wrong with this picture?

For a start, Jeffrey's use of 'true' and 'finally true'. In calling a tenseless proposition 'finally true', he means what most of us would mean by calling it plain 'true'. At any rate, what he calls 'final truth' is what our tenseless judgments aim at, and that is what matters. Given his 'final truth', what he calls 'truth' is entirely immaterial. Suppose I do not know whether the third Test in a (current) 1984 Australian series has finished yet, and so am unsure what tense to give my judgment that England win it. My judgment still has a perfectly definite tenseless content, and attains its object provided England do win, whether they have done so yet or not. That question, whose answer decides whether my judgment is 'true' in Jeffrey's sense, is of no interest to me whatever: 'final truth' is all I am after.

More seriously, suppose that at the end of 1984 I make some tenseless judgment about an event (picked out by a non-temporal description) that happens in a distant galaxy after the light I see left it and before its reflection would return there. If that event is as I judge it to be, my judgment attains its object: it is 'finally true'. Whether, for Jeffrey, it is also 'true' depends on the event's date not being later than 1984, which, according to relativity, may be a matter of an arbitrary choice of reference frame (see Section 3 above): a matter which concerns me not at all, and is certainly not one I can credit with marking the boundaries of objective fact (see Mellor, 1974a; this objection is not met by the modification Jeffrey proposes in his n. 1, p. 259).

I propose to restore 'true' to its customary and proper use, for the intended attribute of all our judgments, tensed and tenseless alike. That is, I shall call 'true' what Jeffrey calls 'finally true'. So I need another term for what he calls 'true'. Since he applies the term to tenseless propositions just when it should be applied to the corresponding past and present tense ones, I shall take the liberty of saying instead that they have 'come to pass'.

I have no objection to Jeffrey's use of 'ineluctable'. By it he means 'necessary', in the sense in which 'it is necessary that p is true if the present state of affairs makes it certain that the p-event will occur, or again if the p-event has already occurred' (Ackrill 1963: 139). The peculiarity of this sense of 'necessary' (in which, for example, p entails its own necessity) is quite enough to justify Jeffrey's preference for 'ineluctable'. It is also what Mackie (1974: ch. 7) means in ascribing 'fixity' to past and present events and the future events they determine. Ackrill and Mackie put the matter in tensed terms, but that, as Jeffrey shows, is by no means essential: an event is 'fixed', we may say, only on and after the date it, or an earlier sufficient cause of it, happens. The tenseless proposition that it happens is likewise 'ineluctable' only on and after the date it, or some other true proposition that determines its truth, 'comes to pass'.

Events therefore, and true tenseless propositions about them, are credited with the ability to undergo at least two sorts of change: (i) the events happen, and the propositions come to pass; and (ii), then or earlier the events become fixed and the propositions ineluctable. What sort of sense can be made of these supposed changes? Tenseless propositions, after all, are normally thought to be unchanging; and while in Section 2 I have admitted that some events *are* changes, I have denied that events themselves change. Nevertheless, sense can be made of (i) and (ii) – only not, as we shall see in Section 7, a sense sufficient for their authors' needs.

Suppose an event e happens at date T. Let H be the property of having happened, and let t and t^* be any dates entirely earlier or later respectively than (every temporal part of) e. Then the change (i) consists in the facts that

$$e \text{ is } \sim H \text{ during } t \tag{9}$$

and

$$e \text{ is } H \text{ during } t^* \tag{10}$$

for all t and t^*.

Do (9) and (10) constitute a change in the sense of Section 2? Certainly, even though e itself is an event and not a thing, (9) and (10) do not reduce to any difference between temporal parts. The parts that would be required, e-during-t and e-during-t^*, are not parts of e, since t is by definition earlier than every temporal part of e, and t^* is later. They would have to be parts of some *ersatz* e-thing, say E, which changes from being

$\sim H$ to being H. But since t is *any* date before e, and t^* any date after it, E would have to span the whole history of the world (except perhaps when e itself is). And in reality there are obviously no such things. An everlasting whole of which World War II-during-5000 B.C., and World War II-during-20,000 A.D. are temporal parts, for example, is not a credible substitute for World War II itself.

So (9) and (10) must be read along the lines of (5) and (6), not (3) and (4): i.e. as

$$\sim H(e, t) \tag{11}$$

and

$$H(e, t^*) \tag{12}$$

H is thus some relation that any event e has to every date later than itself, but lacks to any earlier date. The relation is, of course, a familiar one: 'earlier' is its common name! For an event to 'happen' at a date is simply for it to be earlier than all later dates, and later than all earlier ones.

Now this is not of course a change in e, as it would be were H a real non-temporal relation. Instances of (5) and (6) are indeed taken to imply that a's temporal location includes both t and t^*: it exists at both dates, and at some time between them changes from being G to being G^*. But (11) and (12) imply no such thing about e: on the contrary, they imply that e is *not* located at t and t^*, or it would not be later and earlier respectively than those dates. So e is not an everlasting thing, existing during all the dates t and t^* and changing at T in respect of having happened. Put like that, I dare say no one thinks it is. But there is evidently a recurrent temptation to harbour an equivalent thought: namely, that e's happening is another event, apart from e, and constituting some sort of change in it. Not so: e is all there is, and talk of it happening at T is just a way of saying that T is its date, its temporal location – i.e. that e is later than all times earlier than T and earlier than all later times.

T being e's date is also all there is to the proposition that says this 'coming to pass' (and hence all other true propositions about e doing so). Let p be this true tenseless proposition, and C be the supposed property of having come to pass (i.e. of being 'true' in Jeffrey's eccentric sense). As before, t and t^* are any dates earlier and later respectively than e. Then the facts are that

$$p \text{ is } \sim C \text{ during } t \tag{13}$$

and

$$p \text{ is } C \text{ during } t^* \tag{14}$$

Now Jeffrey in effect construes (13) along the lines of (3) and (4), not (5) and (6); i.e. he credits propositions with temporal parts:

$$\sim C(p\text{-during-}t) \tag{15}$$

and

$$C(p\text{-during-}t^*) \tag{16}$$

Once these temporal parts have come to pass, Jeffrey accumulates them into what he calls 'stages': 'Stages do duty (in the formal mode of speech) for all the facts so far' (253). We may reconstruct his stages from (15) and (16) as follows. For any t (before or after e), let p-through-t be the whole whose temporal parts are p-during-t' for all dates t' containing no instants later than t. Let C^* be the property such that p-through-t is C^* if and only if some temporal part of it is C. Then for any given t, the conjunction of all C^* p-through-t is the stage of the world at t's last instant. In other words, as true tenseless propositions come to pass, they become parts of all later stages of the world.

The mundane facts behind this formal farrago are actually more visible in the relational reading of (13) and (14):

$$\sim C(p, t) \tag{17}$$

and

$$C(p, t^*) \tag{18}$$

Tenseless propositions, unlike events, admittedly have no dates; so C cannot just be the 'earlier' relation, i.e. H. But H suffices to define it:

$$C(p,t^*) =_{\text{df}} H(e,t^*) \tag{19}$$

In other words, e's being earlier than t^* is the fact that makes p have come to pass at that date. p's coming to pass at T, like e's happening then, is in reality nothing more than T being e's date.

So much for (i); what of (ii)? (ii) in fact depends on (i), since fixity depends on events happening, ineluctability on propositions coming to pass. The mere happening of an event fixes it, if the earlier happening of a sufficient cause has not already done so. And no event is fixed until it, or some preceding sufficient cause of it, has happened. Now we have seen that for an event to have happened by a certain date is simply for it to be earlier than that date. The supposed property, H, of having happened is in reality just the 'earlier' relation between events and dates. The supposed property, F, of being fixed is likewise in reality a relation events have to dates: a relation entailed by H but not entailing it, since the earlier happening of a sufficient cause may fix an event before it happens. F is thus definable by H, and by the relation S ($=$ 'is a sufficient cause of'):

$$F(e,t) =_{df} H(e,t) \vee (\exists e^*) [H(e^*, t) \& S(e^*, e)] \qquad (20)$$

As for events, so for propositions. A proposition's coming to pass suffices to make it ineluctable, if it has not already been made so by the earlier coming to pass of a proposition that determines its truth. And no proposition is ineluctable until it, or some such determining proposition, has come to pass. The parallel between propositions and events here is obvious and exact. By definition, p becomes ineluctable just when e becomes fixed: i.e.

$$I(p,t) =_{df} F(e,t) \qquad (21)$$

so the reality of ineluctability is just that of fixity, *viz* the conditions given in (20). All that fixity and ineluctability need are events, their dates, and the tenseless relations 'earlier' and 'sufficient cause'. (And if, as Hume thought, there is in reality no such relation as S, the second disjunct of (20) is always false, and both fixity and ineluctability reduce to events happening, i.e. to their having dates.)

7. FIXITY, COMING TRUE AND TENSE

I have given the simple relational conditions of happening, coming to pass, being fixed and being ineluctable. These conditions are undeniable, but they will hardly satisfy the authors of these conceits. Jeffrey, for example, is trying to conceive the world as 'growing by accretion of facts'. But the reality of his accretion turns out to be nothing more than the truism that the later a date is, the more events are earlier than it. There is no growth in that fact, any more than there is shrinkage in the fact that the earlier a date is, the more events are later than it. Jeffrey must be after something more.

So must Mackie. He hopes to find 'in this notion of fixity a basis for the concept of causal priority' (183). Specifically, causes are distinguished by being fixed at times when their effects are not, but not conversely (180). Since events are fixed at the latest when they happen, this is supposed to explain why causes mostly precede their effects (the exception being later causes fixed before their effects by the still earlier happening of sufficient causes of them). But for this to be an explanation, fixity must not itself be defined by the very fact Mackie wants to derive from it. But in (20) it is. When two causally related events e and e' have no preceding sufficient causes, e is fixed when e' is not just in case e is earlier than e'. So Mackie's definition of causal priority reduces in this case to the cause being the earlier of two causally related events, which is just what he is trying to explain. And when *e* does have sufficient causes, the arbitrary restriction in (20)'s second disjunct, to *e**s earlier than *e*, likewise begs the question it is supposed to answer. Later events, after all, exist no less than earlier ones, and are as capable of being sufficient causes of *e*. If any are, the restriction in (20) discriminates without reason against them; and if none are, it is superfluous.

The fact is that Mackie's theory, like Jeffrey's, is useless and trivial unless having happened and being fixed are something more than the relations I have reduced them to. *H* and *F* must be real non-relational properties of events, acquired at times that are their, or their sufficient causes', dates, for the facts of causal priority to be explained by them. And similarly for *C* and *I*, the coming to pass and becoming ineluctable of Jeffrey's true tenseless propositions. Real accretion must be more than a relational fact: more, at any rate, than the different relations events have to different dates. Can we meet these seemingly modest demands?

Whatever these non-relational properties *H, F, C* and *I* are, their ascription will still have to satisfy the relational conditions I have stated. Maybe 'earlier', as a relation between events and dates, should be defined by 'has happened' rather than *vice versa*: but either way, their equivalence must follow. And even if (19), (20) and (21) will not do as definitions, they must still come out as necessary truths.

What this comes to is that, for example, any judgment to the effect that an event *e* has the property *H* must come out true just in case *e* is not later than the date of the judgment itself. But this is to say that the judgment is indexical: specifically, that its truth conditions are those of the simultaneous judgment that *e* is past or present. In other words, the non-relational property *H* simply *is* that rather imprecise tense: to have happened is to be either past or present.

Ascriptions of fixity are indexical in a slightly more complex way. A judgment that an event *e* has the property *F* is true if and only if its date is not earlier than *e* or some sufficient cause of *e*. For *e* to be fixed, therefore, is just for it, or a sufficient cause of it, to be past or present.

Jeffrey's properties *C* and *I* likewise turn out to depend on tense, despite his tenseless pretensions. If *p* says that *e*'s date is *T*, I judge truly that *p* is *C* if and only if I do so no earlier than *e* itself. So for *p* to have come to pass is for *e* to be past or present. Similarly, for *p* to be ineluctable, either *e* or a sufficient cause of *e* must be past or present.

Mackie and Jeffrey thus both require events to have positions in McTaggart's *A* series, and the changes they postulate are a species of McTaggart change. Events happen and become fixed, propositions come to pass and become ineluctable, as the tense of events changes from future to present. Jeffrey's world growing by accretion of facts is Broad's (1923: ch. II) world growing by accretion of present facts.

We can now, therefore, use the results of Sections 2–5 to extract the truth in Mackie's and Jeffrey's conceits from their error. The truth is that non-relational ascriptions of H, F, C and I, because they are indexical, do not mean the same as non-indexical statements of the relational facts to which I have reduced them. A judgment J, that e is H, is never the same as the simultaneous judgment J', that e is earlier than J. Yet they both have the same truth conditions, namely those stated by J'. And such truth conditions consist entirely of events, including judgments, having dates and being more or less earlier than, or simultaneous with, each other. In the real world that makes these judgments objectively true or false, the non-relational H, F, C and I do not figure at all. Because there is in reality no tense, so there is no real happening of events (apart from the events themselves) and no acquisition of fixity by them; no coming to pass, or becoming ineluctable, of true tenseless propositions.

Fixity, then, since it does not exist, cannot be the real basis of causal priority, nor can the world really grow by accretion of facts. In their intended substance, these conceits will have to go. Still, they will go in good company. Three quarters of a century after McTaggart demolished them, much writing, in many areas of philosophy, still appeals to real, non-relational non-indexical differences between past, present and future. All of that will have to go too. But not from here; despatching so great a multitude of errors must be matter for another place.

REFERENCES

Ackrill, J. L., transl. 1963. *Aristotle: De Interpretatione*. London: Oxford University Press.

Broad, C. D. 1923. *Scientific Thought*. London: Kegan Paul, Trench and Trubner.

Davidson, Donald. 1969. 'The individuation of events.' *Essays in Honor of Carl G. Hempel*, ed. N. Rescher, pp. 216–34. Dordrecht: Reidel.

Geach, P. T. 1979. *Truth, Love and Immortality*. London: Hutchinson.

Goodman, Nelson. 1965. *Fact, Fiction and Forecast*, 2nd edn. New York: Bobbs-Merrill.

Jeffrey, R. C. 1980. 'Coming true.' *Intention and Intentionality*, ed. C. Diamond and J. Teichman, pp. 251–60. London: Harvester.

McArthur, R. P. 1976. *Tense Logic*. Dordrecht: Reidel.

Mackie, J. L. 1974. *The Cement of the Universe*. Oxford: Clarendon Press.

McTaggart, J. McT. E. 1908. 'The unreality of time.' *Mind* 18, 457–84.

McTaggart, J. McT. E. 1927. *The Nature of Existence*, vol. II. Cambridge: Cambridge University Press.

Mellor, D. H. 1974. 'In defense of dispositions.' *Philosophical Review* 83, 157–81.

Mellor, D. H. 1974a. 'Special relativity and present truth.' *Analysis* 34, 74–8.

Mellor, D. H. 1978. 'Conscious belief.' *Proceedings of the Aristotelian Society* 78, 87–101.

Mellor, D. H. 1982. 'The reduction of society.' *Philosophy* 57.

Perry, John. 1979. 'The problem of the essential indexical.' *Nous* 13, 3–21.

Prior, A. N. 1959. 'Thank goodness that's over.' *Philosophy* 34, 12–17.

Russell, B. 1903. *The Principles of Mathematics*. Cambridge: Cambridge University Press.

Williams, Donald C. 1951. 'The myth of passage.' *Journal of Philosophy* 48, 457–72.

Statistical theories, quantum mechanics and the directedness of time[1]

RICHARD HEALEY

A statistical theory, quantum mechanics, occupies a central position in contemporary physics. Perhaps the most commonly drawn metaphysical conclusion has been that determinism is false and indeterminism true. But irrespective of the merits of this conclusion, quantum mechanics prompts the reopening of other metaphysical questions. This paper investigates the bearing of quantum mechanics on some issues in the philosophy of time.

Time seems essentially directed. If there is one property other than dimensionality which appears to distinguish temporal from spatial order, it is this property of directedness. But is the direction of time an objective feature of the world or a subjective feature of our experience of it? Many philosophers as well as scientists have sought a basis for the direction of time in physical processes and the theories we use to understand them. My purpose here is to present and criticise certain attempts to argue that quantum mechanics shows time to be objectively directed.

One can take for granted neither the meaning nor the meaningfulness of the assertion that time is objectively directed. After some preliminary discussion I shall offer two ways of understanding this assertion, and go on to evaluate two associated sets of arguments based on particular features of quantum mechanics for the truth of this assertion as so understood.

I

By way of introduction, consider Reichenbach's attempt in chapter 3 of his book *The Direction of Time* (1956) to distinguish between two properties of time: time may be ordered, or it may be directed. Suppose one were to assert that time has a direction just in case (O) holds.

(O) Under the relation 'later than', time has the order properties of the real numbers under the relation 'greater than'.

[1] I would like to thank the participants at the meeting of the Thyssen Philosophy Group to which an earlier draft of this paper was presented in March 1979 for their helpful comments; Jeremy Butterfield and Huw Price for many fruitful discussions; and most of all Jean Hampton, whose penetrating criticisms helped me not to write a much worse paper.

But (O) is true just in case (O)* is true.

(O)* Under the relation 'earlier than', time has the order properties of the real numbers under the relation 'greater than'.

While (O) or (O)* adequately presses the *ordering* of time, neither captures its *directedness*, for 'we believe that the relation "earlier than" differs structurally from its converse, the relation "later than"' (Reichenbach 1956: 27), and the equivalence of (O) and (O)* points rather to a structural similarity. This prompts a first formulation of the claim that time is objectively directed.

(D) Time is objectively directed just in case there is some structural difference between the relations 'earlier than' and 'later than'.

Now the term 'structural' is unclear. Perhaps the most natural way to clarify it is by generalizing from such obviously structural properties of a relation as reflexivity, transitivity and connectedness, to the following characterization. An open sentence represents a *structural$_L$* property of a relation over a domain just in case the only non-logical terms in that open sentence are predicates designating that relation and the identity relation, and the sentence is satisfied by every sequence of elements of the domain. We can now state (D) more precisely as follows.

(D$_L$) Time is *directed$_L$* just in case there is some structural$_L$ property possessed by only one of the relations 'earlier than' and 'later than'.

Unfortunately, if (O) is true, then (D$_L$) is false. For provided that (O) holds, if 'E' designates the relation of being earlier than, and 'L' designates that of being later than, then there is no open sentence containing only 'E' and the identity predicate such that this sentence, but not the result of substituting 'L' for 'E' uniformly throughout this sentence, is satisfied by an arbitrary sequence of times. The relations 'earlier than' and 'later than' are then structurally$_L$ identical, a conclusion to which the equivalence of (O) and (O)* already pointed.

Faced with the inadequacy of (D$_L$), two alternative approaches to the explication of (D) present themselves. One approach notes a clear difference between earlier and later: the past is earlier than the present, while the future is later. If one focuses on the past and future, then differences seem manifest. For example, we can change the future, but not the past; and we can have records of the past, but not of the future. But if such differences are to be significant here, they must be seen as consequences of some deeper structural difference between the past and the future. Now there is an ancient and influential view of time according to which these and other differences arise from the fact that the future is ontologically open, while the past is settled. It is intuitively clear that on this view time is intrinsically directed, though I postpone further discussion of this intuition to Section 3. These considerations motivate a first way of understanding the assertion that time is directed: time is *directed*$_0$ just in case the future is ontologically open, while the past is settled. It has often been held that time would be directed$_0$ in an indeterministic world, but not in a deterministic world. In Section 3 I further clarify the associated view of time and see how far quantum mechanics shows that time is directed$_0$.

The other approach to the explication of (D) is harder to characterize. The intuition behind it is that time is objectively directed just if the direction of time can be read off from basic physical processes in time, together with the theories in terms of which we understand these processes. For only if this is so can the direction of time have an objective physical basis. There seem to be two features of quantum mechanics which may permit one to read off the direction of time from basic physical processes which obey this theory. If quantum mechanics is not time-reversal invariant, then one may expect basic physical processes to display temporally asymmetric behaviour. And any statistical theory like quantum mechanics treats time asymmetrically in that it is predictive but not retrodictive. Section 2 explains these two features and shows that an improved account of time-reversal invariance collapses them into one. After this preliminary clarification I consider how far this single feature helps one to read off the direction of time from basic physical processes obeying quantum mechanics. The section ends with a critical assessment of this general approach to the directedness of time.

2

The intuition behind one approach to the directedness of time is that time is objectively directed just if the correct temporal orientation can be read off from basic physical processes in time, together with the theories in terms of which we understand these processes. As a first attempt to make this metaphor literal one might suggest that a sufficiently basic *physical* difference counts as a structural difference. Relaxing the definition of a structural$_L$ property in the natural way, one arrives at the following characterization. An open sentence represents a *structural*$_p$ property of a temporal relation just in case the only non-logical terms in that open sentence are predicates designating that relation and the identity relation, and basic physical predicates,[2] and the sentence is satisfied by every sequence of times. This yields the following version of (D).

(D$_p$) Time is *directed*$_p$ just in case there is some structural$_p$ property possessed by only one of the relations 'earlier than' and 'later than'.

But (D$_p$) seems inadequate as a literal rendering of the intuition. Let us say that a description of the contents of some portion of the world over some period of time is *undirected* just if for no pair of events (times) in that period $e_1(t_1)$ and $e_2(t_2)$ does the description entail that $e_1(t_1)$ is later than $e_2(t_2)$. Then time is directed$_p$ just in case there is some temporal asymmetry in a complete but undirected description of the whole of world history in basic physical terms.[3] It is trivially true that time is directed$_p$, as a moment's consideration of any omelette-making episode makes clear. But this does not enable one to assign the correct orientation to time, given only a complete but undirected description of the whole of world history. For nothing in that description entails, for example, that omelette-cooking occurs just after, rather than just before egg-breaking.

It seems then that (D$_p$) formulated a necessary but not a sufficient condition for time to be objectively directed. One cannot read off the correct temporal orientation from the temporally asymmetric but undirected description of the contents of some portion of a world over some period of time, any more than one can tell from a map of an unknown city with no directional indications, which direction on the map is North. Of course, if one knows something about the city, one may be able to do this: for example one may know that all churches in the city face east. Similarly, if one is given additional items of information about the contents of a temporally asymmetric portion of world history, one may be able to tell its overall temporal orientation. But what sort of information will suffice here, and will it be available?

Consider the additional information that each omelette-cooking episode

---

[2] Whatever the list of basic physical predicates may include, it may *not* include explicitly temporal predicates or relations such as 'earlier than' and 'later than'. The actual list would presumably be extracted from the primitive predicates of our ultimate physical theories.

[3] Note here that neither 'earlier than' nor 'later than' is a basic physical predicate. There will be some temporal asymmetry in such a description unless there is a time such that segments of world history temporally equidistant from that time receive exactly parallel but temporally reflected complete basic physical descriptions. I am grateful to Jeremy Butterfield for pointing out to me the need for these clarifications.

occurs just after an egg-breaking episode.[4] If time is directed$_p$ in some portion of world history, then this additional information will suffice to permit the assignment of an overall temporal orientation to that portion, provided two conditions hold. The regularity must be true in that portion of world history; and it must be instantiated (some omelettes must get cooked). But while these conditions may both hold over some weeks in a well-run kitchen, there is little reason to believe that they will hold for an arbitrary portion of world history – consider a night in a poorly run restaurant, or the past second on Jupiter. What seems needed as a source of reliable and applicable additional information is a sufficiently basic temporally asymmetric law of nature, which will hold non-vacuously in an arbitrary portion of world history. If one knew of some portion of world history both that the contents of that portion were temporally asymmetric in accordance with (D_p), and also that they suitably instantiated some temporally asymmetric law of a theory one knew to hold of an arbitrary portion of world history, then one could indeed use this knowledge to assign a temporal orientation to that portion.

Reasoning in this way one may be led to seek temporal asymmetries in the laws of basic physical theories, in the belief that only if such an asymmetry exists is time objectively directed. Some have sought temporal asymmetry in the time-reversal non-invariance of a basic physical theory. H. Mehlberg, for example, says

If all natural laws are time-reversal invariant and no irreversible processes occur in the physical universe, then there is no inherent intrinsically meaningful difference between past and future – just as there is no such difference between 'to the left of' and 'to the right of'. (1969: 363)

He presumably holds that quantum mechanics fails to reveal such an 'intrinsically meaningful difference', since he believes that the theory *is* time-reversal invariant (pp. 370–2). Others have pointed to some temporal asymmetry inherent in the probabilistic character of quantum mechanics. According to John Lucas,

if we adopt a fundamentally probabilistic philosophy of nature, as the success of quantum mechanics suggests we should, . . . we shall not have to think of time as . . . [being] without any natural direction. (1970: 192)

---

[4] Though graphic, this example is in one essential respect misleading. Since neither 'omelette-cooking' nor 'egg-breaking' is itself an undirected description, this information actually presupposes the prior assignment of an overall temporal orientation. It is not difficult to construct an example which is not misleading in this respect. One may give an undirected description of the temporal sequence of positions occupied by a bouncing ball that gradually comes to rest, thus permitting the formulation of a non-question-begging temporally asymmetric regularity concerning bouncing balls.

While, for S. Watanabe,

it is precisely irretrodictability which is related to phenomenal one-wayness. (1965: 56)

   In order to assess such claims it is necessary first to explain what is meant by saying that a probabilistic theory is time-reversal invariant, or that it is irretrodictable. In fact there are two different ways of understanding time-reversal invariance. As conventionally understood, time-reversal non-invariance and irretrodictability turn out to be independent features, either of which a statistical theory may or may not possess. But I shall also develop and defend a less conventional understanding of time-reversal offered by John Earman (1974). On this account, any statistical theory like quantum mechanics is time-reversal non-invariant simply by virtue of being irretrodictable.

   Consider a theory which describes the time-evolution of the state of any system of each of various kinds. A system is in one state at each time. Assume that there is a mapping T which is one-one from the set of states Σ onto itself, and which associates each state $S \in \Sigma$ with its time-reversed state S^T. For a particular theory of this form, the mapping T may be characterized by analogy with features of the states of a system of Newtonian point particles.

   The state of a system of Newtonian point particles may be specified by giving the position and momentum of each particle. The time-reversed state is that state obtained by preserving the position and preserving the magnitude but reversing the direction of the velocity of each particle. One usually says that Newtonian mechanics provides a time-reversal invariant description of the evolution of the state of a system of point particles just in case the (deterministic) law governing the time-evolution of the state of such a system satisfies the following condition: for all states S_a, S_b through which the system passes in its evolution, and all times t_1, t_2, if state S_a at time t_1 evolves into state S_b at time t_2, then state S_b^T at time t_1 would evolve into state S_a^T at time t_2. If the forces governing the evolution of the system depend only on the positions of the particles, and the masses of the particles stay constant, then the description provided by Newtonian mechanics of any evolution of the state of a system of point particles is indeed time-reversal invariant. If a film of the evolution of such a system were run backwards, it would represent behaviour equally consistent with Newtonian mechanics.

   The notion of time-reversal invariance for a statistical theory has to be characterized slightly differently. Suppose the theory Θ issues in statements connecting states at different times which permit one to estimate the chance that a system in one state at one time will evolve into another state at a later time. Write the state at time t as $S(t)$. For $t_1 < t_2$, assume that for each system to which it is applicable Θ has as consequences statistical statements of the form

$$P(S_b(t_2), S_a(t_1)) = p \qquad (t_1 < t_2),$$

one such statement for each pair of states S_a, $S_b \in \Sigma$. This is to be read as follows: 'the probability that the state of the system at t_2 is S_b, given that the state of the system at t_1 is S_a, is p'. One usually says that Θ provides a *time-reversal invariant description* of the evolution of the state of a system just in case these statistical assertions satisfy the following condition: for all states S_a, S_b through which the system passes in its evolution, and all times t_1, t_2 ($t_1 < t_2$), if the system is in state S_a at t_1 and state S_b at t_2, then the probability, given that the state is S_a at t_1, that the state is S_b at t_2, equals the probability, given that the state at t_1 is S_b^T, that the state at t_2 would be S_a^T. Θ provides a time-reversal invariant description of a kind of system just in case it provides a time-reversal invariant description of all evolutions describable within the theory, of the state of any system of that kind. Finally, Θ is *time reversal invariant* just in case it provides a time-reversal invariant description of every kind of system to which it is applicable. Though interesting for certain purposes, the above understanding of time-reversal invariance is potentially misleading in the context of the present discussion. But before indicating why, it is necessary to explain a temporal asymmetry inherent in the statistical character of a whole class of theories.

Consider once more the statistical theory Θ. Suppose $t_1 < t_2$. If one knows the state at t_1, one cannot infer the state at t_2 from Θ: but Θ does permit one to infer, for any state S_b, what the probability is that $S(t_2) = S_b$. On the other hand, if one knows the state at t_2, there is in general no consequence of Θ which permits one, in the absence of further information, to infer $S(t_1)$, or even the probability that $S(t_1) = S_a$. However, one may have additional information which does permit statistical inferences concerning earlier states. Suppose, for each S_c, one knows the prior probability (independent of the knowledge of $S(t_2)$), that $S(t_1) = S_c$: then one may use Bayes' Theorem, together with the knowledge of $S(t_2)$ and the theory Θ, to infer a revised probability for $S(t_1)$ as follows:

$$P(S_a(t_1), S_b(t_2)) = \frac{P(S_b(t_2), S_a(t_1)) \cdot P(S_a(t_1))}{\Sigma_c[P(S_b(t_2), S_c(t_1)) \cdot P(S_c(t_1))]} \, .$$

Θ treats time asymmetrically, since it issues directly in conditional probabilities for states at later times, not at earlier times. Following Watanabe (1965) I will call Θ a *predictive* but not a *retrodictive* statistical theory, and summarize the temporal asymmetry inherent in the fact that

Θ is predictive and not retrodictive by calling Θ *irretrodictable*. It is important to note that one cannot rectify this asymmetry by expanding Θ by the addition of a further set of statistical consequences of the form:

$$P(S_a(t_1), S_b(t_2)) = p' \qquad (t_1 < t_2).$$

One cannot do this because these reverse conditional probabilities must satisfy certain constraints in order to preserve consistency with the original forward conditional probabilities. For example, for $t_1 < t_2$, if I_1, I_2 are two states the system may be in at t_1, and F_1, F_2 are two states the system may be in at t_2, we must have (in general):

$$\frac{P(I_1, F_1) \cdot P(I_2, F_2)}{P(I_1, F_2) \cdot P(I_2, F_1)} = \frac{P(F_1, I_1) \cdot P(F_2, I_2)}{P(F_2, I_1) \cdot P(F_1, I_2)}$$

where the conditional probabilities on the left are the newly introduced reverse probabilities, and the conditional probabilities on the right are the original forward probabilities. And if one decided to add to Θ reverse conditional probabilities which did meet all constraints, then Θ would change its character radically. For a complete set of non-trivial forward conditional probabilities together with a complete set of reverse conditional probabilities determine a unique set of *unconditional* probabilities of all states at both t_1 and t_2. In this case the statistical laws of Θ would no longer be restricted to laws governing the time evolution of states: Θ would then be imposing *lawlike* boundary conditions on the system! Thus this temporal asymmetry of Θ is by no means an arbitrary feature of the theory. It is moreover an asymmetry with no analogue in a non-statistical theory.

Notice that on the above understanding, consistent with its irretrodictability, Θ may or may not be time-reversal invariant; while time-reversal invariance has not even been defined for a retrodictive theory. Now this is strange, because if the time-reversal operation has been appropriately defined, one would expect that if a predictive theory is true of a set of systems, then a retrodictive theory should be true of the time-reversed states of these systems. Indeed the operation called time-reversal above is more appropriately referred to as a *reversal of motion* (Earman 1974: 25–6). Henceforth, I shall rename all time-reversal properties of theories defined above as the corresponding motion-reversal property. A genuine time-reversal operation, applied to the evolution of a system, would reverse not the motion of the system in time, but rather the overall orientation of time. Thus one should say that Newtonian mechanics provides a time-reversal invariant description of the state of a system of point particles just in case for all states S_a, S_b through which the system passes in its evolution, and all times t_1, t_2, if state S_a at t_1 evolves into state S_b at t_2, then under the opposite temporal orientation state S_b^T at time t_2 evolves into state S_a^T at time t_1.

Under very general conditions, a deterministic theory is motion-reversal invariant just in case it is time-reversal invariant, so the confusion of these two notions is relatively harmless. But for a statistical theory the distinction is crucial; for while an irretrodictable theory may or may not be motion-reversal invariant, it is never time-reversal invariant: time-reversal invariance would require that a statistical theory be both predictive and retrodictive. On the revised account, a statistical theory provides a time-reversal invariant description of the evolution of the state of a system just in case the following condition holds: for all states S_a, S_b through which the system passes in its evolution, and all times t_1, $t_2(t_1 < t_2)$, if the system is in state S_a at t_1 and S_b at t_2, then the probability, given that the state is S_a at t_1, that the state is S_b at t_2, equals the probability, under the opposite temporal orientation, that given that the state at t_2 is S_b^T, the state at t_1 is S_a^T. But no law of an irretrodictable theory will specify the latter probability, and so no irretrodictable theory can be time-reversal invariant.

Though statistical, quantum mechanics differs from the simple theory Θ considered above in several ways, and these differences require certain modifications in applying the above notions of motion-reversal, time-reversal, and irretrodictability to quantum mechanics. I treat these subtleties in the appendix, and consider there whether the theory does or does not possess each of these properties. But suppose for now that quantum mechanics had the structure of theory Θ. Then quantum mechanics would be irretrodictable and so time-reversal non-invariant: and it may or may not also be motion-reversal invariant. How could any of these asymmetries help one to assign a temporal orientation to an arbitrary portion of world history?

If one knew that theory Θ was true, then one would certainly be justified in expecting that the frequencies of occurrence of various sequences of states in any sufficiently large portion of world history would approximately reflect the probabilities specified by Θ. For it is then very likely that for any sufficiently large portion of world history there is an assignment of temporal orientation to that portion such that this is so. But there is no reason to expect that for *both* possible assignments of temporal orientation the frequencies of sequences of states reflect the probabilities specified by Θ. The irretrodictability of Θ thus makes reasonable the following procedure for assigning a temporal orientation to a sufficiently large portion of world history. Pick an arbitrary temporal orientation: if under that orientation the frequencies of sequences of states approximately reflect the probabilities of Θ, assign that temporal orientation; otherwise, assign the opposite temporal orientation.

Does the reasonableness of this procedure for assigning a temporal orientation show that time is objectively directed? There are several immediate problems with this inference. First, even if Θ were a basic physical theory, there would be no guarantee that an arbitrary portion of world history contains any systems to which Θ applies, and for this reason the procedure will not invariably lead to the assignment of the correct temporal orientation to an arbitrary portion of world history. Second, it is consistent with the truth of Θ that for a few portions of world history the frequencies of sequences of states do not approximate to the probabilities of Θ. Third, it may be that in a very few portions of world history the frequencies of sequences of states coincidentally approximate to the probabilities of Θ for *both* temporal orientations. For the last two reasons, in an arbitrary portion of world history in which the laws of Θ are instantiated the procedure is not guaranteed to lead to the correct assignment of temporal orientation. Notice that since irretrodictability implies time-reversal non-invariance, these problems also undermine the claim that the time-reversal non-invariance of Θ shows time to be objectively directed. I postpone for a paragraph the assessment of the seriousness of these problems.

The motion-reversal non-invariance of Θ for all states of all systems in that portion would be neither a necessary nor a sufficient condition for the success of the above procedure in assigning the correct temporal orientation to a portion of world history. Even if such a portion contains enough examples of motion-reversal non-invariant state sequences to yield frequencies which for one temporal orientation approximate convincingly to many motion-reversal non-invariant laws of Θ, this in no way guarantees that this is not also true for the opposite temporal orientation. And even if a portion contains enough examples of motion-reversal invariant state sequences to yield frequencies which for one temporal orientation approximate to many motion-reversal invariant laws of Θ, this in no way guarantees that this is also true for the opposite temporal orientation.

We have seen that if one knows that an irretrodictable statistical theory like Θ is true then one has a reasonable procedure for assigning a temporal orientation to a sufficiently large portion of world history, but that there are problems in inferring from this fact that time is objectively directed. It is now time to investigate the basis of this approach more closely. Recall that (D_p) gave an inadequate sense to the claim that time is objectively directed since it alone provided no basis for the assignment of the correct temporal orientation to a portion of world history. Knowledge of the truth of an irretrodictable theory like Θ appears to go some way toward remedying this defect. But is this knowledge legitimately available to one? It is not if all legitimate knowledge must be obtained from observation of the contents of world history. For if one knows the temporal orientation of these contents one does not need to know that a theory like Θ is true in order to orient temporally any portion of world history; while if one does not know the temporal orientation of these contents, how does one come to know that Θ rather than its time-reversed Θ^T is true? There is a vicious circularity involved in arriving at the truth of Θ observationally, then using its truth to orient world history temporally.

The circularity is less blatant if one arrives at the truth of Θ from observations in portion A of world history, then uses the truth of Θ to orient temporally some non-overlapping portion B of world history. This procedure is open to the epistemological objection: how can observations in A guarantee that Θ is true in B? But the objection is merely a variety of scepticism about induction, and there are more pressing objections. As discussed above, the procedure is not guaranteed to succeed in assigning the correct temporal orientation to B. But this means that it cannot be used as a criterion of whether or not time is objectively directed in B. If it could be used in this way, then there would be no sense to the claim that in some instance the procedure failed to assign the correct temporal orientation to B. Since this does make sense, there is clearly some other criterion for assigning a temporal orientation to B with which the results of this procedure usually coincide, but to which the results of this procedure are actually subordinate. Moreover, in so far as this other criterion assigns a temporal orientation to every portion of world history, it appears to guarantee that time is objectively directed, irrespective of any procedure based on Θ.

It is clear what this criterion is: we somehow orient temporally some local portion of world history, and then the order properties of time determine the temporal orientation of every other portion of world history. It seems then that time is objectively directed just in case the local temporal orientation procedure is objective, and time order is objective. Now time order is objective in so far as our most successful physical theories are set in a spacetime background which guarantees the existence of such a time order. In particular, non-relativistic quantum mechanics is set in a Newtonian spacetime background, which implies that (O) holds.

Is there an objective basis for any local temporal orientation procedure? I shall pursue this question further in Section 4, but conclude the present section by examining one attempt to argue for an affirmative answer by appeal to our knowledge of the truth in our local portion of world history of some irretrodictable theory (e.g. Θ).

Suppose examination of a temporally undirected description of the contents of our local portion of world history revealed that under one overall temporal orientation an irretrodictable theory like Θ was true. Then under the opposite temporal orientation, the time-reversed theory Θ^T would be true. Θ^T is retrodictive but not predictive. If one could argue on *a priori* grounds that in these circumstances Θ rather than Θ^T must be the correct theory for our local portion of world history, then one could assign to that portion the temporal orientation that renders Θ true of it. And one might then argue that the contents of our local portion of world history, as given by an undirected description, provide an objective basis for this local temporal orientation procedure.

Such an *a priori* argument may be based on a pragmatic preference for an irretrodictable theory over a theory that is retrodictive but not predictive. We require a predictive theory in order to be able to predict the future state of a system when we know its present state. We require such a theory because we do not now know the future state. And the theory aids us in controlling processes because we can prepare a system so that it has some prearranged present state, and we can thus control at least the probability that it will have some particular future state. On the other hand, we do not need a theory to tell us about past states of a system, since in principle we could always have records of these past states. And we have no interest in attempting to control past states of a system, because we know that any such attempt is futile, as the past is already settled and unchangeable. Clearly then a theory which is only predictive is pragmatically preferable to one which is only retrodictive. Thus, given only that either Θ or Θ^T is true of our local portion of world history, but without knowledge of the overall temporal orientation of that portion, we can legitimately infer that the correct temporal orientation is that which renders Θ, not Θ^T, true locally. For we know in advance that the true theory must be predictive.

This argument is interesting but unconvincing. If we are dealing with a fundamental statistical theory whose probabilities reflect the way the world is at the deepest level, then it is wilful optimism to assume that these probabilities will have conveniently arranged themselves so that a predictive not a retrodictive theory correctly describes them. It may be convenient for us if the true theory is predictive, and one may even be able to argue that we could not in fact ever stumble across the truth of a fundamental retrodictive theory. But such pragmatic considerations cannot suffice to show that Θ, not Θ^T, is *correct* as well as preferable, and therefore fail to warrant the objectivity of the associated temporal orientation procedure.

I conclude that even if we did know that quantum mechanics is a true irretrodictable theory, the connection (if any) between this fact and the claim that time is objectively directed must be much less direct than was suggested by the metaphor with which this section began.

3

Consider now the attempt to explicate the sense in which there is a structural difference between 'earlier than' and 'later than' by appeal to the fact that the future is open while the past is settled. This gives rise to the following formulation of the directedness of time.

(D_0) Time is *directed*$_0$ just in case the future is ontologically open, while the past is settled.

The view that time is directed$_0$ may be traced back at least to Aristotle; and it is interesting that different commentators on Aristotle have attributed to him each of the two variants of this view which must now be distinguished.[5]

One who asserts the ontological openness of the future may be making a modal claim or a claim about the indeterminateness of actual truth-values. He may be asserting that many possible future courses of events are equally compatible with everything that has occurred until now: or he may be asserting that there are statements about the future which are presently neither true nor false. William James (1917) linked indeterminism to the modal openness of the future, and determinism to its denial. According to 'our ordinary unsophisticated view of things', he says, 'actualities seem to float in a wider sea of possibilities from out of which they are chosen; and *somewhere*, indeterminism says, such possibilities exist, and form part of the truth. Determinism, on the contrary, says they exist *nowhere*, and that necessity on the one hand and impossibility on the other are the sole categories of the real' (150–1). C. D. Broad (1923) gives a particularly clear statement of the alternative way of understanding the ontological openness of the future, and explicitly links this view to the direction of time (though not to indeterminism). His theory, he says, 'accepts the reality of the present and the past, but holds that the future is simply nothing at all . . . The sum total of existence is always increasing, and it is this which gives the time series a sense as well as an order' (66). And 'judgements which profess to be about the future do not refer to any fact . . . at the time they are made. They are therefore at that time neither true nor false. They will become true or false when there is a fact for them to

---

[5] The key passage from Aristotle appears in chapter ix of *De Interpretatione*. Donald Williams (1951) reads this passage as proposing the truth-value gap account of the openness of the future. Elizabeth Anscombe (1964) provides her own translation and elucidation of the passage, according to which Aristotle rather advances the modal account.

refer to; after this they will remain true or false, as the case may be, for ever and ever' (73).

These two explications of the ontological openness of the future may be called the *modal* account and the *truth-value gap* account. Suppose one can show in either of these two senses that the future is ontologically open, while the past is settled. How does this relate to the claim that time is objectively directed in the sense of (D)? The structural difference between the relations 'earlier than' and 'later than' is manifested by the fact that there is a wide class of pairs of open sentences, such that neither sentence in each pair is true of any time later than the present, while exactly one sentence in each pair is true of any time earlier than or simultaneous with the present. These sentences may be modal, in which case a typical pair would be $<\Box F(t), \Box \sim F(t)>$, where $F(t)$ ascribes some state to a system at time t: or they may be non-modal, in which case a typical pair would be $<F(t), \sim F(t)>$ with the same notation. In this way, if one could show in either the modal or the truth-value gap sense that the future is ontologically open while the past is settled, this would indeed justify the claim that time is objectively directed. My purpose here is not to decide the issue of the correctness of either of these accounts but rather to investigate how the assumption that quantum mechanics is a true fundamental theory bears on this issue, and specifically on the question as to whether time is directed$_0$.

Consider first the truth-value gap account of the openness of the future. Perhaps the most promising argument for this account from the assumed truth of quantum mechanics as a fundamental theory is verificationist. If quantum mechanics is true, then there is no way even in principle that we could presently know the truth-value of certain statements about the future. For an old-fashioned verificationist, this would imply that such statements are (presently) meaningless, and hence lack truth-value. A contemporary anti-realist might argue rather that the present epistemic inaccessibility of at least some aspects of the future implies that the only notions of present truth and falsity which we could have learnt, and whose use our linguistic behaviour could manifest, are not such as to guarantee that each statement about these aspects of the future is now determinately true or false. Suppose then that a verificationist construes the truth of statements about the future as follows: a statement about the future is true just in case there is now a situation whose existence we can now acknowledge as justifying the ascription to that statement of the value true.[6] It will follow that if there is nothing about the present situation

---

[6] I have adapted this formulation from an analogous principle concerning the past given by Michael Dummett in his paper 'The Reality of the Past' (1978: 368).

which justifies the relevant truth-ascriptions, then there are statements about the future such that neither they nor their negations are presently true. The law of excluded middle will fail for such statements. In an early paper, 'The Causal Structure of the World', Reichenbach appears to link indeterminism to the truth-value gap account of the openness of the future in just this way. He says there: 'If events are not totally determined, it cannot be claimed that the future is already established . . . The past, on the other hand, is definite, and the present is that threshold over which the universe steps in going from an indefinite state to a definite one.' (Reichenbach and Cohen 1978: vol. 2, p. 87.) And his later paper 'The Logical Foundations of Quantum Mechanics' (*op. cit.*, pp. 237–78) has been taken to argue that the indeterminism of quantum mechanics shows, via this link, that the truth-value gap account of the future is correct.[7]

But if one considers the verificationist's motivations, there are reasons to query one step in his argument that the future is open in the sense that there are truth-value gaps. The argument begins by noting that if quantum mechanics is true there is no way even in principle that we could presently know the truth-value of certain statements about the future. This is then supposed to ground a fundamental distinction between these statements about the future and any statements about the present or past, namely all statements about the present or past now possess determinate truth-values, while these statements about the future do not. Consistent with verificationism, this distinction is to be made by appealing to what truth-values we could in principle presently know. But is it correct to say that we could presently know the truth-value of any statement about the present or past?

The modality involved in this claim requires clarification. Taken in one way, the claim that we could presently know the truth-value of any statement about the present or past, but of no statement about the future, follows immediately from the truth-value gap account of the future, and so may be used to explain that view by drawing out an epistemological consequence. In this way of taking it, one could (in principle) know to be true just what in fact *is* true. But if one does take the modality in this way, then it is quite unclear how to seek support for the truth-value gap account of the future by arguing from quantum mechanics to the truth of the above epistemic claim.

If one is to give such an argument, the modality had better be taken as a variety of *physical* possibility. But in what sense is it physically

---

[7] Adolph Grünbaum (1964: 319–21) at least takes this to be his argument. However, it seems to me questionable whether this *is* Reichenbach's argument here. In particular, Grünbaum's translation of 'indéterminisme' as 'indeterminateness' on p. 320 makes one suspect an over-enthusiastic attempt to read the argument into the text.

possible presently to know the *present* state of any system? Presumably knowledge acquisition relies on physical processes, and it is physically possible to come to know that X just in case there is some physical process which may serve as a basis for the acquisition of knowledge that X. Now in order to determine whether or not this is so, a physical characterization not only of such processes but also of the knower is required: what is physically possible for you may not be physically possible for me. Perhaps one can assume some particular physical characterization of a 'standard knower'. But if such a standard knower is located within some finite spatial region at time t, then in order that he, she or it may come by knowledge of the state at t of some system S outside that region, it must be physically possible for there to be processes which, at t, convey to the standard knower correct information about the state of S at t. Any such process must involve instantaneous action at a distance, and this is not physically possible on any currently plausible physical theory.

Suppose that this last point is granted, and that it is admitted that strictly speaking it is only physically possible presently to know the state of any distant system at some time in the past (perhaps the very recent past). There still appears to be a substantial difference between the standard knower's capacity for acquiring knowledge about the future state of a system, according as such states are related deterministically or indeterministically to a state at some time in the past such that the knower could (physically) have now come to know this earlier state. For if these states are related deterministically, then it seems that the knower could come to know the future state merely by coming to know the past state, while if they are related indeterministically this would not be so. But there is only an apparent difference here between an indeterministic and a deterministic theory. Even in the case of a deterministic theory, the knower is only able to use knowledge of the past state to acquire knowledge of the future state if he, she or it also knows that the system remains undisturbed by outside influences in the intervening period. But while it may be reasonable to assume that this will be so, it cannot presently be *known* to be so. In order to come to know the future state of a system through use of a deterministic theory, the standard knower would have to know not just the state of that system, but also enough about the past states of all other systems to determine that that system will remain isolated in the intervening period. But in any reasonable physical theory it will be physically impossible for the knower presently to have that knowledge: the knowledge will be available only after the intervening period is over, at which time it becomes superfluous.

In summary, the verificationist may seek to appeal to the indeterministic character of quantum mechanics in order to show that certain statements ascribing a future state to a system lack present truth-value, and that in this sense the future is ontologically open. But it is not clear that the argument distinguishes between present and future states in the way it is intended to. And more importantly, the argument turns out not to depend on the indeterministic character of quantum mechanics after all: when spelt out, it is clear that this argument may be made with just as much (or as little) force on the basis of a deterministic theory like classical mechanics.

At this point the proponent of the truth-value gap account may modify his tactics. Instead of seeking an argument for the *truth* of his account on the basis of quantum mechanics, he may welcome quantum mechanics as displacing a deterministic theory (classical mechanics), on the basis of which an argument could be given to the *falsity* of his account. For if a deterministic theory were true, and statements about the past and present had determinate present truth-values, then statements about the future would also have to have determinate present truth-values; while the abandonment of a deterministic theory for an indeterministic theory seems to sever this link. However this distinction between a deterministic and an indeterministic theory also turns out to be merely apparent. For far from permitting the truth-value gap account of the future, quantum mechanics as customarily formulated is *also* inconsistent with it.

Any physical theory, like quantum mechanics, which describes the temporal evolution of the state of one of a general class of systems is blind to such indexical features as presentness. Whether or not they are statistical, the laws of such a theory are expressed in such a way as to concern only the (quantitative) earlier/later than relations between certain events and processes, without regard to whether these events and processes occur now, in the past, or in the future. One may specify independently of these laws how some such events and processes are temporally related to the present, but this specification cannot affect the truth-conditions of sentences of the theory. Thus, independently of the details of how theoretical sentences acquire truth-conditions, the truth-conditions of a sentence referring to a time t must be given independently of whether t is now present, past or future. To say that a physical theory is true is to say that the truth-conditions for sentences in the language of the theory are such as to render true those sentences expressing laws of the theory. And to say that a law of time-evolution is true is to say that the truth-conditions of sentences describing earlier and later states of systems are such as to ensure that the truth-conditions for the law obtain.

Suppose one argues on the basis of an indeterministic theory that a sentence describing the state of a system at a time lacks truth-value if and only if that time is in the future. Now consider many instances of a statistical law whose antecedents refer to times that are now past or present, but whose consequents refer to times that are now future, each attributing some state to a system. None of the consequents is true, since each concerns a future time. But then each instance of the statistical law is a conditional whose antecedent is true but whose consequent is not true. If we assume that these sentences obey classical logic, then none of them is true. In this way one may amass enough counter-instances to any statistical law of time-evolution linking states of these types with non-zero probability to justify its rejection, though not of course to falsify it outright.

The only way to maintain the truth of any statistical theory in the face of such putative counter-examples to its time-evolution laws would be to change the logic in which the theory is formulated. To maintain that a sentence ascribing a state to a system at a time has determinate truth-value just in case that time is present or past is inconsistent with the acceptance as true of a physical theory formulated within classical logic which includes laws governing the time-evolution of these states, whether the laws are deterministic or statistical. In so far as quantum mechanics is such a theory, that time is directed$_0$ according to the truth-value gap account is not merely not supported by the assumed truth of quantum mechanics: it is inconsistent with it.[8]

The modal account of the ontological openness of the future is not similarly inconsistent with quantum mechanics. But how can the assumed truth of quantum mechanics be used to argue that the future is open in this modal sense, and will this show that time is directed$_0$? According to the modal account, the present state of the world is compatible with more than one future course of events: all these alternative futures are possible, though only one will be actual. The truth of an indeterministic theory is taken to make room for such an account whereas a deterministic theory would exclude it by holding only a single future course of events to be compatible with the present state of the world. The claim then is that quantum mechanics, as an indeterministic theory, implies that time is directed$_0$, since if it is true there are many possible future courses of events though only a single past course of events.

Assume for the moment that it is true that quantum mechanics implies laws which assign non-zero probability to more than one later state, given any particular earlier state, of a system. It then follows that more than one future course of events is compatible with the present state of the world.

<hr>

[8] Various non-classical logics have been proposed for quantum mechanics, and one may wonder whether the adoption of one of these would enable one consistently to maintain the truth-value gap account of the openness of the future. I think the answer is 'no', but to argue this here would produce an unreasonably lengthy digression.

But in order to establish that time is directed, one must also show that only a single past course of events is compatible with the present state of the world. One cannot show this solely by appeal to quantum mechanics. As a predictive but irretrodictable theory, quantum mechanics directly places *no* restrictions, even statistical ones, on what past course of events is compatible with the present state of the world. Indeed, for quantum mechanics the past seems even more open in a modal sense than does the future! Thus quantum mechanics does not show time to be directed in the modal sense; it is merely not inconsistent with this claim in the way a deterministic theory would be. The intuition that the past is *not* modally open remains. But in order to show that time is directed$_0$ it will be necessary to defend this intuition against the objection that it originates not from any structural feature of time itself, but merely from certain pervasive aspects of our situation as knowers and agents, such as our ability to have records of the past but not of the future, and our ability to affect the future but not the past.

As is discussed more fully in the appendix, quantum mechanics as most frequently interpreted does *not* imply laws which simply assign non-zero probability to more than one later state, given any particular earlier state of a system. Instead, the theory yields statistical predictions as to the subsequent state of a system if (and only if) a measurement is performed on that system, where a measurement is a particular type of irreversible physical interaction. Now no law of quantum mechanics has anything to say about whether a measurement will or will not take place on a system at some particular later time. But in order to use the theory to make any estimate at all as to the future behaviour of a system, one needs to know details of what measurements will occur, and when they will occur. This is not a practical problem, since we can decide what measurements we shall make, and when we shall make them. But it is philosophically interesting, because it seems to make room for human intervention (of a rather abstruse and colourless sort) in the workings of the universe unrestricted by physical law. To modify a famous dictum of Einstein, it seems to be we who are playing dice with the world, not God. Actually, if measurement is thought of as a purely physical process, then it seems reasonable to suppose that this process could occur whether or not we arrange for it to do so. And the natural reaction to the apparent gap in the quantum mechanical description of the world is to seek a supplementary theory to fill it. Attempts to fill this gap by further applications of quantum mechanics itself are highly problematic.[9] But in any event, what this gap seems to indicate is a deficiency in our present theorizing about the world,

---

[9] One way to view the quantum measurement problem is as a barrier to such attempts.

not an unsuspected falling short of the scope of physical law. The regular and controllable nature of measurement processes supports this interpretation.

The status of measurements in the statistical laws of quantum mechanics does not affect the conclusion that quantum mechanics fails to show that time is directed$_0$ in the modal sense. Though there is a way in which the future seems even more open in this theory than in a theory with simple indeterministic laws relating earlier and later states, this does nothing to show that the past is not open if quantum mechanics is true. However, there is a different argument for the openness of the future in the truth-value gap sense which does depend on this peculiar feature of the statistical laws of quantum mechanics. A crude version of this argument goes as follows. We know that the future is open because we can freely act so as to affect it. But if all statements about the future now had determinate truth-value, then nothing we could do could affect the future. Hence the future is ontologically open in the sense that there are truth-value gaps.

As it stands this latest argument is not a good one. Perhaps it can be embellished so as to seem more convincing. But my concern is not with the details of the argument but with the question of how quantum mechanics bears on the premise: that we are able freely to act so as to affect future events. For it may be thought that it is just this premise that is incompatible with the truth of a fundamental deterministic physical theory, yet compatible with the truth of quantum mechanics. The traditional response to this last thought is to point out that motion in accordance with statistical physical laws is no less (and no more) incompatible with free action than is motion in accordance with deterministic physical laws. But the peculiarities of quantum mechanics as a statistical theory may seem to raise a new possibility. For the statistical laws of quantum mechanics place no restrictions, not even statistical ones, on what measurements occur and when they occur. Perhaps then the traditional response may be evaded by locating our ability to act freely in our freedom to perform quantum measurements, unconstrained in this activity by the laws of quantum mechanics?

Apart from the fact that, as pointed out above, we seem to be faced here with a deficiency of physical theorizing rather than with a falling short of the scope of physical law, there are further objections to this libertarian counter. Everyone apparently performs free actions, but only experimental physicists (occasionally, and with difficulty) perform quantum measurements: why then is free action so easy if it involves the performance of quantum measurements? This is an invalid objection: clearly any action no matter how trivial involves physical processes of which the agent is quite unaware. A better objection is this: if free action is made possible by the freedom to perform quantum measurements, then on what system is such a measurement performed, and what is the physical nature of the system that performs the measurement? More importantly, in order to be able to control whatever system it is that performs the quantum measurement one must be able to act freely: but then another quantum measurement is needed to make possible the first, and we have the start of an infinite regress.

The peculiarities of measurement in quantum mechanics may affect the details of the debate as to the compatibility of free action with the truth of a fundamental statistical theory, but the structure of the debate remains the same. Indeed, this particular debate seems rather insensitive even to important differences in the kind of physical theory one may suppose to be true of the world, including its human denizens. And even given the truth of the premise that we are able freely to act so as to affect future events, it is no easy matter to construct a good argument for the ontological openness of the future, or for the directedness$_0$ of time.

4

The last two sections failed to unearth any convincing reasons why quantum mechanics should be taken to show time to be objectively directed. It is worth pointing out that though quantum mechanics prompts the re-opening of this and other metaphysical questions, there is a general reason why it is unlikely to lead to their resolution. The reason is that no metaphysical argument based on actual features of quantum mechanics can be expected to carry conviction while there continues to be a heated debate about the correct interpretation of even the best-established and formally most perspicacious expression of that theory. The unquestionable success of quantum mechanics during the past fifty years has done little to resolve the dispute over its correct interpretation which began with the theory. As the appendix shows in greater detail, metaphysical conclusions from quantum mechanics are highly interpretation-dependent; for instance, there are interpretations which deny that quantum mechanics is indeterministic. Until one has an agreed interpretation of quantum mechanics, any metaphysical conclusions from that theory are best thought to be merely conditional in form, with little confidence placed in the truth of the antecedent!

I conclude with some brief remarks on the broader question of the bearing of physical theory in general on the issue as to whether or not time is objectively directed. The first remark is this: in investigating the implications of physical theory for our concept of time it is surely perverse not to concentrate attention on spacetime theories. If directedness is an objective physical property of time itself, then surely this must be a consequence of a theory of spacetime structure (relativistic or non-relativistic) rather than of any particular class of processes set in that structure? From this perspective, quantum mechanics hardly seems a promising place to look for the direction of time, set as it is within a wholly familiar spacetime structure. But neither Newtonian nor (at least special-) relativistic spacetimes ascribe any property of directedness to time.

Is one to conclude that in so far as one's background spacetime theory provides no basis for distinguishing 'earlier than' from 'later than', time has no objective direction? Recall from Section 2 that any temporal orientation procedure based on a particular theory of physical processes in spacetime is subordinate to a procedure which somehow fixes the temporal orientation locally and then extends this to distant regions of spacetime via the temporal order properties of that spacetime as specified within the theory appropriate to it. Provided that the spacetime is temporally orientable, this procedure leads to the assignment of a consistent temporal orientation throughout the spacetime. But is this procedure objective?

It may seem that it is not objective, since any local temporal orientation procedure will be acceptable only if it assigns the 'earlier than' relation to paradigmatic pairs of nearby events which are simply *perceived* to bear that relation. And thus the basis of the fundamental temporal orientation procedure seems inherently subjective. We immediately experience one event as later than another, and this is what underlies both our notion that time is directed and also, apparently, our fundamental criterion for assigning a temporal orientation to events. But note that this criterion is not subjective, but rather *inter*subjective: we all agree on the assignment of temporal orientation in our local neighbourhood.

But intersubjectivity is not yet objectivity: we may all agree that a statement is true even though it is in fact false. Could we all be wrong about the direction of time? If not, does our intersubjective agreement not fail to guarantee objectivity? I can see no simple sense in which we could all be wrong about the direction of time; but in spite of this I think that time is objectively directed. An analogy may help here: we all (certain people, including the colour-blind, excepted) agree on which things are paradigmatically red. We could not all be wrong in this agreement, but still these things are *objectively* red. An important reason for this objectivity is that we possess at least a sketch of a causal explanation for our intersubjective agreement which appeals to properties of red things, the effect of the reflection of light from their surfaces to our eyes, etc. In just the same way we possess a sketch of a causal explanation for our intersubjective agreement as to the sequence of paradigmatic nearby events.

The existence of this explanation is one factor which goes to bolster up our intersubjective agreement on time direction into the objective directedness of time. And it is here that physical theory bears on this objective directedness. To see how, consider certain elementary facts that underlie our intersubjective agreement on time direction. We could hardly agree on the sequence of nearby events without memory, and the ability to act in the world and to communicate with one another are also arguably required. The appropriate task for one who wishes to argue that there is a physical basis for the objective directedness of time is to investigate the physical basis of these abilities. One may be able to show that it is the pervasive thermodynamic asymmetry in our local region of the universe that permits the physical processes which underlie these abilities. In so far as this demonstration is successful, it reveals a temporal physical asymmetry which in an indirect way provides a foundation for the objective directedness of time. But the search for this kind of foundation would be a very different enterprise from those which have received criticism in earlier sections of this paper. I see no reason to expect quantum mechanics to be relevant to this search.

APPENDIX

The body of the paper, with two exceptions, assumed that quantum mechanics is a statistical theory with the structure of theory Θ. The appendix removes this simplifying but false assumption. This is necessary, as it is only because quantum mechanics itself (in some form) constitutes a fundamental and well-entrenched component of contemporary physical theory, that this paper has relevance to arguments that time in fact *is* objectively directed. I begin by presenting a skeletal account of the statistical character of standard non-relativistic quantum mechanics in accordance with a mainstream interpretation of that theory. There follows a discussion based on this account of the notions of time-reversal, motion-reversal and irretrodictability as applied to quantum mechanics. The appendix concludes with some remarks on the consequence of adopting one of a number of alternative interpretations of the theory.

Quantum mechanics is not one theory but many. I shall restrict attention to the conventional Hilbert space formulation of non-relativistic quantum mechanics as developed by Dirac (1930) and given mathematical rigour by von Neumann (1932). It is worth remarking that fundamental as this theory undoubtedly is to contemporary physics, it is not considered to provide a true account of the basic structure of the physical world.

However, its formal properties are relatively well understood, and it serves as a model for the development of successor theories which seek to improve on its known deficiencies. Thus it may be reasonable to expect a true physical theory to share significant features with the theory to which I restrict attention here, if it is reasonable to talk in this strongly realist way at all.

The first respect in which quantum mechanics differs from theory Θ is in its more complex notion of state. One may attempt to specify the state of a system at a time by giving the numerical values of some set of quantities ('observables') on the system. But while there is a traditionally recognized full set of observables on a system, the state is *not* specified by giving the value of each observable in this set: indeed, it is held never to be the case for any system that all these observables possess values simultaneously. Rather, there exist many subsets of the full set of observables such that all the observables in any such subset may possess values simultaneously, while some other observables outside this temporarily privileged subset do not have values at all at that time.

There is another notion of state which ties up with such state specifications via the values of a certain subset of observables. This is the state vector (or state function, or wave function). The state vector at a time yields probabilistic predictions as to the value obtained in a measurement of any observable on the system at that time. In addition, there is an equation (the time-dependent Schrödinger equation, or TDSE for short) which, if a system is isolated and if no measurement is performed on it during some time interval Int, uniquely relates the state vector at time t_1 to the state vector at time t_2, for any times t_1, t_2 in Int. The connection between the two notions of state is this: the state vector at a time assigns unit probability to some unique value as the result of a measurement of each observable in some maximal subset of observables at that time.[10] It therefore seems natural to suppose that each observable in this set has that value to which the state vector assigns probability unity. Then the state vector at a time is uniquely correlated with the state as specified by giving the values of a maximal subset of observables at that time, and we may regard the two notions as different components of a single notion of state.

If we make this identification an interesting consequence follows. The state of a system changes in two different ways. If the system is isolated and no measurement is performed on it, the state evolves continuously and deterministically in accordance with the TDSE. If the system is subjected to a measurement, then its state changes discontinuously and

---

[10] I ignore the complications involved in treating an observable with a continuous part to its spectrum.

statistically in accordance with a statistical prescription provided by the nature of the measurement process and the state vector of the system immediately prior to the measurement.[11] On this interpretation, quantum mechanics is a deterministic theory except when a measurement occurs, but the occurrence of measurement introduces an irreducibly statistical element into the theory's treatment of the temporal evolution of the state of a system.[12] I assume for the moment that measurement is some irreversible physical process.

The theory issues in non-statistical laws of the form:

(1) If the state of a system at t_1 is S_a and no measurement occurs in the interval $[t_1, t_2]$, then the state at t_2 is S_b. Symbolically, $S_a(t_1) \mathrel{\&} \sim M_{[t_1, t_2]} \rightarrow S_b(t_2)$.

Note that in (1) t_2 may be later than, earlier than, or (trivially) simultaneous with t_1. The theory also issues in statistical laws of the form:

(2) If the state of a system at $t - \varepsilon$ is S_a and a measurement of type α occurs at t, then the probability that the state at $t + \varepsilon$ is S_b is p. $P(S_b(t + \varepsilon), S_a(t - \varepsilon) \mathrel{\&} M_t^\alpha) = p$.

Combining laws of these two forms one obtains more interesting statistical laws of the form:

(3) $P(S_b(t_3), S_a(t_1) \mathrel{\&} \sim M_{[t_1, t_2)} \mathrel{\&} M_{t_2}^\alpha \mathrel{\&} \sim M_{(t_2, t_3]}) = p$.

Note that in (2) ε is a positive infinitesimal, and in (3) $t_3 > t_2 > t_1$. Note also that in (2) and (3) the idealization that measurement is instantaneous has been made. It is not, however, necessary to assume that all measurements are ideal in that they obey the projection postulate: instead, each measurement is classified as of a specific type α, such that (2) holds for measurements of type α. Ideal measurements then give special cases of (2) and (3).

Despite the fact that quantum mechanics as here presented gives rise to three distinct forms of time-evolution law, the form (3) may be taken as fundamental. For (1) is not statistical at all, and yet the crucial aspect of quantum mechanics in this context is its statistical character: and (2) is a relatively uninteresting special case of (3).

Statistical laws of the form (3) differ in important respects from the simple statistical laws of the form $P(S_b(t_2), S_a(t_1)) = p$ which occurred in theory Θ. These laws no longer simply relate states at different times, of a

---

[11] I ignore the intermediate mixed state on the controversial but simplifying assumption that it merely represents our ignorance of the actual (pure) state after the measurement.

[12] I ignore the further (major) problems of non-separability concerning the relation between the states of a system before and after a non-measurement interaction.

RICHARD HEALEY

system which is left to itself in the interim. Rather, they relate states before and after the occurrence of a measurement on the system. How does this difference affect the discussion of time-reversal, motion-reversal and irretrodictability based on the simpler statistical laws of theory Θ?

Consider first the question of motion-reversal invariance. If the state of an isolated system is represented by S, then one may define a motion-reversed state S^T and compare the restrictions on the time-evolution of S^T to those imposed on the time-evolution of S. The co-ordinate space representative of the state vector of a spinless quantum system is the wave function $\psi(r_1, r_2, \ldots)$. With the above understanding of the state of a quantum system, the state is the ordered pair of the state vector ψ (or its representative in some basis) and a sequence of values of observables in the maximal set of observables assigned unit probability by ψ. The analogy with classical mechanics then requires that the time-reversed state be the ordered pair with first member $\psi^*(r_1, r_2, \ldots)$, and second member a sequence of values of the observables in the maximal set assigned unit probability by $\psi^*(r_1, r_2, \ldots)$. This guarantees, for example, that if the state S is such that the total momentum is p, then the value of the total momentum in state S^T is $-p$; while if the position of the centre of mass of the system in S is r, then the position of the centre of mass in S^T is also r.

If this is how we are to understand the motion-reversed state of a (simple) quantum system, is quantum mechanics motion-reversal invariant? Restricting attention to a law of the form (1) one may wish to give an affirmative answer, since $(1)^T$ is a law if (1) is.

$$(1)^T \quad S_b^T(t_1) \ \& \sim M_{[t_1,t_2]} \rightarrow S_a^T(t_2)$$

This expresses the familiar fact that the TDSE is invariant under 'time-reversal' (i.e. what we are now calling motion-reversal). But it does not follow that quantum mechanics is motion-reversal invariant, for one must also consider laws of the form (3), and here matters are less clear, since no notion of motion-reversal invariance has yet been defined for a statistical law of this form. One may say that a theory incorporating laws of this form is motion-reversal invariant just in case it gives a description of every kind of system to which it is applicable which is motion-reversal invariant in the following sense: for all states S_a, S_b through which the system passes in its evolution, all times t_1, t_2, t_3 ($t_1 < t_2 < t_3$), and all types of measurement α, $(3)^*$ is a law if (3) is.

$$(3)^* \quad P(S_a^T(t_3), S_b^T(t_1) \ \& \sim M_{[t_1,t_2)} \ \& \ M_{t_2}^\alpha \ \& \sim M_{(t_2,t_3]}) = p$$

In this sense quantum mechanics is *not* motion-reversal invariant. But this explication may be criticized on the grounds that an α-type measurement should not be referred to in the correct motion-reversed law. And indeed, if one restricts attention to a certain class of ideal measurements, it is possible to define a natural mapping $M^a \rightarrow M^{*a}$ such that (3)** is a law just in case (3) is.

$$(3)** \quad P(S_a^T(t_3), S_b^T(t_1) \;\&\; {\sim}M_{[t_1, t_2)} \;\&\; M_{t_2}^{*a} \;\&\; {\sim}M_{(t_2, t_3]}) = p$$

The measurement M^a is an ideal measurement of the observable P such that

$$P \cdot U_{t_2 t_3}^{-1}(\psi_b) = \lambda_p \cdot U_{t_2 t_3}^{-1}(\psi_b)$$

and the measurement M^{*a} is an ideal measurement of the quantity Q such that

$$Q \cdot U_{t_2 t_3}^{-1}(\psi_a^*) = \lambda_q \cdot U_{t_2 t_3}^{-1}(\psi_a^*),$$

where $U_{t_2 t_3}$ is the time-evolution operator for the interval $(t_{t_2, t_3}]$.

Fortunately it is not necessary here to decide the issue of the correct understanding (if any) of motion-reversal for quantum mechanics, in view of the irrelevance of this decision to any philosophically interesting issue concerning the directedness of time (cf. Section 2 of this paper).

On the natural understanding of irretrodictability, quantum mechanics is irretrodictable in so far as it issues in statistical laws of the form of (3) but in none of the form (3)#,

$$(3)\# \quad P(S_a(t_1), S_b(t_3) \;\&\; {\sim}M_{[t_1, t_2)} \;\&\; M_{t_2}^a \;\&\; {\sim}M_{(t_2, t_3]}) = p.$$

Without going into details, it is clear that this is sufficient to show that quantum mechanics is not time-reversal invariant, since, however, this last notion is explicated, a time-reversal invariant statistical theory must be required to generate some reverse (as well as forward) probabilities; that is, probabilities for an earlier state, given a particular later state. But since quantum mechanics is irretrodictable it does not do this.

The adoption of a different interpretation of quantum mechanics may change more or less radically one's view of how the theory treats time. I now briefly consider some alternative interpretations. There are two features of quantum mechanics as I have sketched it which have seemed unsatisfactory. Some have objected to the irreducibly statistical character of the theory, and have sought to supplement it by further 'hidden variables' which would permit this statistical character to be understood after the manner of classical statistical mechanics, so that the laws of time-evolution for the revised notion of state become deterministic though not time-reversal invariant. Relocating the origin of the statistical nature of quantum mechanics in this way, if successful, would make quantum mechanics even less like theory Θ. But there are reasons to believe that this approach will not be successful, since hidden-variable theories have been shown to have the physically objectionable feature of non-locality. However, it has been suggested that non-locality would not appear so physically objectionable if a limited amount of what seems appropriately described as backwards causation were to occur over very short time intervals (Costa de Beauregard 1977). If this suggestion were both coherent and ultimately convincing, then this would certainly be an interesting metaphysical consequence of the treatment of time in quantum mechanics.

Many people have objected to the presence of talk of measurement in the laws of quantum mechanics. Wigner (1967) has suggested that this talk is ineliminable, and that measurement is a process whereby the mind (not the brain) acts directly on the quantum state of a system. There are severe problems with this interpretation even within physics. Should it nevertheless prove tenable, then again quantum mechanics would have interesting metaphysical implications relating to time. The objections considered in Section 3 to the idea that free action requires quantum measurements would have to be reconsidered, since the ultimate measuring system would then be an otherwise non-physical experiencing ego. And there would be a close link between time asymmetry and the direction of time, in so far as the experiencing ego both has immediate awareness of the direction of time in its successive experiences, and also gives rise to a basic asymmetry of physical processes in time by performing the quantum measurements which are the source of this asymmetry. But Wigner's suggestion is perhaps more intriguing than coherent.

Other interpretations try to eliminate talk of measurement altogether. One way is by modifying the TDSE so that it becomes possible to use quantum mechanics itself to eliminate this talk. Such an attempt may well affect the time-reversibility of the description of the temporal evolution of an isolated quantum system. But even this type of approach must apparently in addition incorporate some statistical law of time-evolution, if it is to avoid the introduction of hidden variables.

There is one interpretation which claims to give a fully deterministic and time-reversal invariant account of the state of the universe, which account nowhere refers to measurements. This is the Everett interpretation (DeWitt and Graham 1973). Irreversibility and talk of measurements enter not into the basic account of the evolution of the universe, but only into the account of our observation of this evolution. In this interpretation also it may seem that it is we who supply both the asymmetry and the direction of time. I suspect that this is not the intention of the interpretation's proponents. But since I also suspect that the interpretation is ultimately incoherent, I shall not pursue it further here.

REFERENCES

Ackrill, J. L., transl. 1963. *Aristotle: Categories and De Interpretatione*. Oxford: Clarendon Press.
Anscombe, G. E. M. 1964. 'Aristotle and the Sea Battle.' Smart (1964), pp. 43–57.
Broad, C. D. 1923. *Scientific Thought*. London: Kegan Paul, Trench and Trubner.
Čapek, M. A. 1961. *The Philosophical Impact of Contemporary Physics*. New York: Van Nostrand Reinhold.
Costa de Beauregard, O. 1977. 'Two Lectures on the Direction of Time.' *Synthèse* 35, 129–54.
Davies, P. C. W. 1974. *The Physics of Time Asymmetry*. London: Surrey University Press.
DeWitt, B. and Graham, N. 1973. *The Many Worlds Interpretation of Quantum Mechanics*. Princeton, N.J.: Princeton University Press.
Dirac, P. A. M. 1930. *The Principles of Quantum Mechanics*. Oxford: Clarendon Press.
Dummett, M. 1978. *Truth and Other Enigmas*. London: Duckworth.
Earman, J. 1974. 'An Attempt to Add a Little Direction to "The Problem of the Direction of Time".' *Philosophy of Science* 41, 15–47.
Grünbaum, A. 1964. *Philosophical Problems of Space and Time*, 1st edn. New York: Knopf.
Grünbaum, A. 1970. 'The Meaning of Time.' *Essays in Honor of Carl G. Hempel*, ed. N. Rescher. Dordrecht: Reidel.
James, W. 1917. 'The Dilemma of Determinism.' *The Will to Believe and Other Essays in Popular Philosophy*. London: Longmans Green.
Lucas, J. R. 1970. *The Concept of Probability*. Oxford: Clarendon Press.
Mehlberg, H. 1969. 'Philosophical Aspects of Physical Time.' *The Monist* 53, 340–84.
Reichenbach, H. 1956. *The Direction of Time*. Berkeley: University of California Press.
Reichenbach, M. and Cohen, R. S., eds. 1978. *H. Reichenbach: Selected Writings, 1909–1953*, vol. 2. Dordrecht: Reidel.
Sklar, L. 1974. *Space, Time and Spacetime*. Berkeley: University of California Press.
Smart, J. J. C., ed., 1964. *Problems of Space and Time*. New York: Macmillan.
von Neumann, J. 1932. *Mathematical Foundations of Quantum Mechanics*, transl. R. T. Beyer. Princeton, N.J.: Princeton University Press, 1955.
Watanabe, S. 1965. 'Conditional Probability in Physics.' *Supplement of the Progress in Theoretical Physics*, vols. 33–4; pp. 135–60.
Wigner, E. 1967. *Symmetries and Reflections*. Cambridge, Mass.: M.I.T. Press.
Williams, D. 1951. 'The Sea Fight Tomorrow.' *Structure, Meaning and Method*, ed. P. Henle *et al.* New York: Liberal Arts Press.

Time, reality and relativity

LAWRENCE SKLAR

I

There is a doctrine, venerable and very familiar, that that which does not exist in the present does not, properly speaking, exist at all. Alternatively there is the equally ancient and equally intuitive view that only the past and present have determinate reality and that the future has no such being, or at least no such determinate being.

For the moment I don't wish to explore the fundamental questions to which these doctrines give rise: 'Why do they have the intuitive appeal they do?', 'Can any good reasons whatever be given to support them?', 'In the final analysis could they possibly be correct?' Rather I want to take a look at what purports to be a simple and conclusive refutation of all such doctrines. For, it has been claimed, 'science' refutes the asymmetric treatment of the present and non-present once and for all. While this argument too has been 'floating around' for some time, it has fairly recently appeared in the literature, once (Rietdijk 1966) in a version replete with infelicities of expression and formulation (talking about determinism of the future when it is the question of determinate reality which is at issue) and the other time (Putnam 1967) framed with greater philosophical sophistication. But in both cases the argument is fundamentally the same.

Consider an observer at a place-time. According to the doctrine in question, events in his future (say) are not determinately real. But according to relativity there is going to be another observer, coincident with the first, and hence certainly real to him, since immediately present to him. Now many of the events future to the first observer will be present to the second, so long as the two observers are in relative motion. Indeed, for any future event (relative to the first observer and spacelike separated from him) there will be a second observer such that that event is present to the second observer when the second observer is coincident with the first. So the 'future' event will be real, relative to the second observer. But surely 'being real to' is a transitive notion. If the event is real to the second observer who is real to the first, it must be real to the first observer, contradicting our original claim that events future to an observer lack reality for him.

Once we have accepted the principle of transitivity of reality then we can go further. For even events in my future light cone will be present, hence real, for some observer who is present, hence real, for an observer in motion with respect to me but coincident to me and, hence, real for me. So even events in my absolute future must be declared to have determinate reality.

Now obviously the whole argument rests upon the fundamental assumption of the transitivity of 'reality to'. Given the relativization of simultaneity to a reference frame in relativity, anyone who wishes to relate determinate reality to temporal presence must also relativize having reality to a state of motion of an observer. And given the non-transitivity of simultaneity in relativity across observers in differing reference frames, we could also easily find our way out of this argument by simply denying that 'having reality for' is a transitive relation. Now Putnam calls the transitivity of 'reality for' the principle of there being 'No Privileged Observers', and, surely, we would like all observers to have equal rights to a legitimate world-description (1967: 241). But why one would think that such a doctrine of 'No Privileged Observers' would lead one immediately to affirm the transitivity of 'reality for', given that one has already relativized such previously non-relative notions as that of simultaneity, is beyond me.

But simply blocking the argument against the traditional doctrine in this way is of very little interest. For example we still are at a loss as to what specific doctrine about reality we should adopt in the relativistic case. Should we simply relativize the old doctrine taking, as before, the present to be the real and simply denying the transitivity of 'reality for'? Or should we adopt some alternative, more radical, view? To decide requires that we look a little more closely into the metaphysical presuppositions which underlie the relativistic spacetime picture itself. What I want to do here, rather than pay attention only to the problem as so far narrowly construed, is to explore the more general issue suggested by Putnam in the concluding remarks to his paper (1967: 247):

I conclude that the problem of reality and determinateness of future events is now solved. Moreover, it is solved by physics and not by philosophy . . . Indeed, I do not believe that there are any longer any *philosophical* problems about Time; there is only the physical problem of determining the exact physical geometry of the four-dimensional continuum that we inhabit.

I think that such a naive view is as wrong as can be. Just as a computer is only as good as its programmer ('Garbage in, garbage out'), one can extract only so much metaphysics from a physical theory as one puts in.

Time, reality and relativity

While our total world-view must, of course, be consistent with our best available scientific theories, it is a great mistake to read off a metaphysics superficially from the theory's overt appearance, and an even graver mistake to neglect the fact that metaphysical presuppositions have gone into the formulation of the theory, as it is usually framed, in the first place.

2

The original Einstein papers on special relativity are founded, as is well known, on a verificationist critique of earlier theories. Referring to the observational facts, generalizable as the null-results of the generalized round-trip experiments, he argues for the necessity of finding an 'operational' meaning to apply to simultaneity for events at a distance and for the impossibility of doing this in any way which allows us an empirical determination of the one-way velocity of light. From then on the moves are all well known which invoke the 'radar' method for establishing simultaneity, this resting upon the conventional stipulation of the uniformity of the velocity of light in all directions. Nothing in the way of newly predicted phenomena, such as those predicted by the mechanics designed to preserve the old conservation rules relative to the newly adopted spacetime picture, changes the fundamental point – that quantities are to be introduced into one's theory only insofar as they are empirically determinable or conventionally definable from empirically determinable quantities.

Of course, even within the relativistic context it is easy to forget the verificationist arguments which initiated the theory in the first place, and to forget the distinction within the foundations of the theory between the genuinely factual elements and the merely conventional choices which go into the relativistic spacetime picture. Hence the necessity for frequently re-introducing the claim of conventionality and the difficulty of seeing it through to its full consequences even for those who espouse it. Thus we have the Reichenbach–Grünbaum argument to the effect that in picking ε equal to $1/2$ in the familiar 'radar' definition of simultaneity, a merely conventional element has been introduced, corresponding to picking the velocity of outgoing and returning light as equal. But it then takes some effort to see that even the choice of a linear relationship on the time of emission and reception of a light signal to define the point simultaneous with the light's being reflected at a distant point is itself undetermined by the facts. Thus the very choice of a flat spacetime for the spacetime appropriate to special relativity is easily argued to be as much a 'mere conventional choice' as is that of the equal velocities of light in opposite directions.

Now it might be argued that Einstein's verificationism was a misfortune, to be encountered not with a rejection of special relativity, but with an acceptance of the theory now to be understood on better epistemological grounds. There is precedent for an attitude of this kind. Einstein was led to general relativity both by an attempt to satisfy Mach's requirements for an explanation of inertial forces and by the belief that covariance of equations represented a generalization of the relativity principle underlying special relativity. Most would now take the theory itself to be our best available current theory of gravitation but would deny its conformity with Machian requirements and would deny the legitimacy of identifying covariance with a relativity principle of any kind. Even the principle of equivalence, another 'background principle' to general relativity, is questionable in a way the theory is not.

But I don't think a position of this kind will work in the present case. I can see no way of rejecting the older aether-compensatory theories, originally invoked to explain the Michelson–Morley results, without invoking a verificationist critique of some kind or other. And I know of no way to defend the move to a relativised notion of simultaneity, so essential for special relativity, without first offering a critique, in the same vein as Einstein's, of the pre-relativistic absolutist notion, and then continuing to observe that even the relativistic replacement for this older notion is itself, insofar as it outruns the 'hard data' of experiment, infested with a high degree of conventionality.

3

Once adopt a verificationist stance of any kind and certain fundamental questions arise. First of all there is the question of just what properties and relations are going to be taken as epistemically available by 'direct inspection'.

Obviously one such relation is that of coincidence between events. Without taking this as epistemically available to us, the whole project of providing an 'operational definition' for simultaneity for distant events is blocked at the start. For we must be able to synchronize clocks at a point (determine the emission and reception time of the light beam, etc.) and this amounts to a determination of spatiotemporal coincidence. It is less frequently noted that another fundamental notion is taken as primitive in the definition as well. This is the notion of continuity along a causal (timelike or lightlike) path. The definition requires the use of the times of emission and reception of a reflected light beam. But how could these be determined unless there were available to us some method of determining that the light beam whose later reception time is determined is the very same light beam whose emission time was earlier recorded and whose reflection at the distant point coincided with the event at a distance which is to be identified as simultaneous with some local event? And if, as seems the only plausible move to make in this case, we take identity of the beam through time (or genidentity of the set of events making up the beam) to be constituted by the spatiotemporal continuity of the beam (and what else would serve in its place?), this presupposes that a determination of such continuity is epistemically available to us.

Now there are many absolutely crucial questions to be asked here. Can any coherent sense be made of the claim that our spatiotemporal knowledge is exhausted by reference to data formalizable in the two notions of coincidence and causal continuity? Can an intersubjective physical theory be formulated in terms of such 'immediately accessible' concepts at all? To what extent is the restriction of epistemic accessibility to such a limited class of notions in any sense a commitment to a denial of 'reality' to those features of the world whose description outruns the capacity of the meager basis to which we have limited ourselves? We will return to some of these questions shortly when we ask ourselves what the consequences are for a doctrine of the 'irreality of the past and future' of such an epistemic foundation for relativity.

For the moment, however, I want to digress a little to show how in a slightly different, but closely related, context the natural choice of these two concepts as primitive once more suggests itself.

Even more primitive as a feature of our spacetime than metric features, like simultaneity, is its topology. Both the Minkowski spacetime of special relativity and the pseudo-Riemannian spacetime of general relativity have determinate topologies – the usual manifold topologies. To what extent are these topologies empirically determinable and to what extent, once again, are they merely conventionally stipulated? A natural suggestion has been to look for the source of our epistemic access to topology in the causal structure among the events in the spacetime.

What we soon discover is this. In Minkowski spacetime there is, indeed, a topology definable solely in terms of the causal connectivity among events, the Alexandrov topology, which is provably equivalent to the ordinary manifold topology. Furthermore, the Alexandrov topology is 'natural' enough that one might be inclined to the view that what we really meant all along by the topological structure of the spacetime was that which the Alexandrov topology gives us. Essentially, it identifies the basis of open sets of the topology with sets of events timelike accessible from a pair of events, i.e. an open set in the basis is the common region of the interior of a forward light cone from one event and the interior of the backward light cone from another.

When we go to the general relativistic picture, it is plain that there are pathological spacetimes which are such that the Alexandrov topology and the manifold topology will not even be extensionally equivalent. Only if the spacetimes are what is called strongly causal will the two topologies coincide. Indeed, it is easy to show that no topology defined in terms of causal connectibility in any way will be generally adequate even to capture extensionally the usual manifold topology of the spacetime.

A recent result of Malament's (1977a) does show us, however, that if one assumes the spacetime to have some manifold topology or other, then the topology along the causal paths (timelike and lightlike paths) does completely determine the topology of the spacetime, in the sense that any one-to-one mapping from one spacetime to another which preserves continuity along causal paths will be, relative to the assumption that both spacetimes have the usual manifold topology, a homeomorphism.

But why assume the spacetimes have a manifold topology of the usual sort? Are there any other topologies which one can imagine which will differ from the manifold topologies but which will agree with them regarding the structure of continuity along causal paths? There are. For example, the topology which takes the maximum number of sets as open sets (the finest topology on the spacetime) compatible with the continuity structure along causal paths will agree with the usual manifold topology along these paths and yet be non-homeomorphic to the manifold topology.

Now why are any of these mathematical results of any philosophical interest? Clearly the usual epistemic presuppositions are being made: in order to attribute a structure to the world we must indicate how that structure can be empirically determined by us. The full topological structure of spacetime is not the sort of thing open to any kind of direct or immediate epistemic access. But the continuity structure along causal (lightlike or timelike) worldlines is. Hence the topological structure of the world is determined only to the extent that it is fixed by the continuity structure of the causal worldlines. And why would one think continuity along causal worldlines open to us? Only, I think, because of the implicit claim that they can be 'lived along' by observers who, in their traversal of the worldlines, could directly determine the continuity or discontinuity of a set of points (Sklar 1977).

4

Let us now return to our original question: to what extent is the old doctrine of the unreality of the past and future undermined by the adoption of a relativistic spacetime picture of the world? Alternatively, can a metaphysics be constructed which retains the old 'intuition' and which is compatible with the new spacetime view? And if the latter is the truth, what modifications in the old view must we make in order to retain this compatibility? And how plausible, given the original standpoint of the 'irrealist' metaphysics, are these modifications going to be?

Now one obvious way to reconcile irrealism with relativity is simply to drop the principle of the transitivity of 'reality for', and retain the original doctrine in the form most similar to its original version, that is, taking as real for an observer all and only those events which are temporally present to him. Of course we must now relativize 'reality for' so that it is just as non-transitive across observers in mutual motion with respect to one another as is simultaneity, but there doesn't seem to be anything very objectionable *a priori* about this. Making this move certainly doesn't seem to be positing any observer as 'privileged', as Putnam alleges when he calls the principle of transitivity of reality the principle that There are No Privileged Observers. For, just as with simultaneity in special relativity, all inertial observers are on a par and none are singled out as in any way 'privileged'.

But there does seem to be *something* wrong with this approach. The source of our skepticism about it lies, I think, in the strong pressures toward a conventionalist, and, hence, in a sense, irrealist, theory with regard to simultaneity for distant events. If the totality of our epistemic access to the theory is contained in the facts about coincidence and continuity along causal paths, can we reasonably take a realistic attitude toward relations not totally reducible in terms of these notions at all? Of course we can *call* two spacelike separated events simultaneous or not, picking the relativistic definition for distant simultaneity or some other. We can speak of light as having an isotropically distributed velocity, or not, again choosing either the standard relativistic convention or some other. But if any one of these accounts, framed in superficially inconsistent terms, can explain equally well all the hard data of experience, why should we take the accounts as differing at all in the real features they attribute to the world? We are easily led to the (standard) conventionalist claim that there is no fact of the matter at all about the equality of the velocity of light in all directions, and no fact of the matter at all about which distant events are 'really' simultaneous with a given event.

If we now associate the real (for an observer) with the simultaneous for him, we must, accepting the conventionality of simultaneity, accept as well a conventionalist theory of 'reality for'. It is then merely a matter of arbitrary stipulation that one distant event rather than another is taken as real for an observer. Now there is nothing inconsistent or otherwise formally objectionable about such a relativized notion of 'reality for', but it does seem to take the metaphysical heart out of the old claim that the present had genuine reality and the past and future lacked it. For what counts as the present is only a matter of arbitrary choice, and so then is what is taken as real. At this point one can easily see why one would adopt, instead, the line standardly taken by proponents of relativistic spacetime and declare all events, past, present and future, equally real. For the distinction among them, being reduced to a mere conventional way of speaking, seems far too fragile to bear any real metaphysical weight.

5

But there is an alternative. It is a radical one and one we will hesitate to take when we see the position into which we are being forced. It is, however, not only consistent with the relativistic spacetime picture, but an option well in keeping, from at least one point of view, with at least some of the original motivations underlying the irrealist view about past and future.

Given an observer at a time, what, from the relativistic point of view, should we cast into the domain of unreality? Certainly all the contents of his absolute past and future, that is everything contained in either his forward or backward light cones.

But what about the contents of the world outside of both light cones altogether? The first alternative we looked at discriminated among these events, taking those simultaneous (in the relativistic sense) with the observer at the instant to be real and discarding the remainder as unreal. But such a discrimination seems all too arbitrary. The obvious alternatives are to count everything spacelike separated from the observer as real or to count it all as unreal for him. And surely of these two alternatives it is the latter which best suits the initial motivation behind the irrealist viewpoint as originally construed.

Not that the former option has no plausibility. After all we do sometimes speak of the region outside the lightcones as the region of the 'absolutely simultaneous', and if reality is to be identified with temporal presentness this does suggest that all which is now 'absolutely present' should be counted in the domain of reality. I don't think there is any way in which one could 'refute' this option. We are, after all, revising the older theory of the reality of only the present to fit the new world picture and, without some further constraints on our choices, I suppose the options are up to us.

Nonetheless this view certainly seems peculiar. Having dismissed as unreal things whose only deficiency is the fact that causal signals from them have taken time to arrive at us now, or that causal signals from us will take some time to arrive at them, it seems very suspicious indeed to promote into the domain of the fully real those things causally inaccessible to us (now) altogether.

But just what were some of the motivations behind the original intuition of the irreality of all but the present? Perhaps if we get some grip on them we will have some guidance in our attempt to rework the theory to fit these changed circumstances.

(1) One source of the old intuition is plainly the fact that our natural language is tensed. We speak of things now as existing but of the past only as having existed and of the future only as going to exist. But, of course, we can't look here for an explanation of the intuition of unreality, except possibly in a weak psychological sense. For the claim of the irrealist is that this natural way of speaking is not a mere trivial artifact of ordinary language but reflective of some deep metaphysical distinction between the elsewhen and the now. What we would like is some interesting distinction between past and future and the present which simultaneously explains the felt metaphysical distinction and its representation by means of tense distinctions in ordinary language.

(2) There is the resort to irrealism which holds that an irrealist view of the future is necessary to avoid fatalism. This connects the alleged irreality of the future with an alleged present absence of truth-values for future tensed statements. The claim is made that this absence of truth-values for future tensed statements allows us to avoid the disturbing conclusion that what will be is already the case, and, in the sense that statements about it already have a truth-value, is somehow fated. It allows us to avoid thinking that the future is as beyond our capabilities of changing as are facts about the past.

From this standpoint we obviously want events in the forward light cone to be unreal. From this perspective we will, contrary to the general time-irrealist point of view, want events in the backward light cone to be real, as sentences about them presumably have, now, determinate truth-value. But what are we to do with events outside either light cone?

By our previous argument from the arbitrariness of distant simultaneity, we will probably not want to discriminate among the events at spacelike separation some we take to be real and some unreal. So our only option would be to take all of them as either real or all as not real. Given that irreality, from this motivational standpoint, was invoked in order to prevent there already being a fact of the matter which, now, determined the truth or falsehood of future tensed assertions, there seems to be little to constrain our choice of reality or unreality for those events which are now absent from both our causal past and our causal future.

Suppose we take all events at spacelike separation to be unreal. We will, of course, have to suffer some consequences which, from the 'ordinary language' point of view are rather peculiar. For example there will be events which are now such that they will be in my real past at some future time, but which will never have a present reality to me at all (Putnam 1967: 246). But we expect that a move to a relativistic picture will force some violence on our ways of speaking and this is no refutation of adopting this way of thinking about things. Nor, from this point of view (the sole aim of which is to prevent the over-abundant reality of our to-be-experienced future) do I see any reason why we couldn't adopt the other alternative of simply taking all spacelike separated events as now real. Of course both of these options will have the virtue (if it is one) that two coincident observers, no matter what their relative motion, will, at an instant, attribute reality and irreality to the same regions of spacetime.

(3) The irrealism which we are primarily interested in is not, however, that which attributes reality to past and present and denies it only to the future, but rather that which takes the real to be only that which is present. What is the motivation behind this view, over and above the, possibly merely artifactual, special role of tensed discourse in our natural language?

At least one motivation for the view is to be found in the 'epistemic remoteness' of past and future. This ties in the familiar intuition about the irreality of past and future with such familiar verificationist themes as the claim that all propositions about past and future are, unlike those about the present (or, at least, about present immediate experience) 'inferential' in nature. And it ties it in with the further move, so familiar with radical verificationist programs, either to reduce statements about past and future to statements about present evidence or, alternatively, to adopt some kind of 'criterion' theory of the meaning of statements about past and future, taking their meaning to be fixed by their relation, in terms of warranted assertability, to statements about present experience which are exhaustive of the body of evidential statements for them. From the latter point of view it is fairly clear what the assertion of irreality to past and future amounts to (denial of bivalence to past and future tensed statements, etc.) and from the former it is at least clear why a radical asymmetry of some sort is being maintained between, on the one hand, present tensed statements and, on the other, those tensed past or future.

I certainly don't intend to examine either of these familiar verification-ist claims, either in general or in their particular application to statements about past and future. Rather, I want only to explore what the impact of relativistic spacetime thinking ought to be on the programs. Let me also say here that to the objection that it is not these verificationist themes which really underlie our intuitions about the asymmetry of reality in question, I have no reply. I think the objection may be correct, but I am at a loss to imagine what other source to the asymmetry there might be (over and above, of course, the mere fact of asymmetric grammatical forms in our particular natural language).

Now looking at the asymmetry as presently motivated the first thing to say, once again, is that there is nothing in the relativistic picture to force the asymmetrist to give up his position. But how shall he frame it so as to fit the new spacetime picture? Once again he could simply relativize everything to the state of motion of an observer, using the relativized simultaneity notion to demarcate present from past and future. But, as usual, this is subject to the objection of arbitrariness and, in addition, from the point of view of our present, epistemic, motivation for the asymmetrical standpoint we have a far more plausible option to take.

If the past and future are to be declared unreal due to their 'epistemic distance' from us, what attitude are we to take toward events at spacelike separation from us? The answer is clear. For events at spacelike separation from us are now (although they may not be in the future) totally immune from epistemic contact by us. That is the very fact which in special relativity leads us to the doctrine of conventionality of simultaneity in the first place, and in the context of general relativity leads to the notions of event and particle horizons, deterministic but unpredictable spacetimes, etc.

So surely in this case the option is fixed for us. If we are to take the past and future as unreal due to their epistemic distance from us, surely we are to declare everything outside the lightcone as unreal as well.[1]

What is the most interesting here is this: even from the pre-relativistic point of view, it isn't the least bit clear why we should have ever treated the elsewhere differently than the elsewhen. At least it isn't clear until we have some backing for the intuition of unreality of past and future which doesn't rest upon the mere fact of epistemic distance. For even from the pre-relativistic point of view, events spatially separated from us seem to be as epistemically distant as those in the past or future. Why, even in the pre-relativistic situation, would anyone have wanted to deny reality to the

[1] There is also the position that the past, being now forever epistemically inaccessible to us, is unreal, but that the present and the future are real as both are, now, open to present or future inspection. From the point of view I have been proposing I think it evident what the relativistic revision of this doctrine ought to be. Surely the regions outside the light cone are to be lumped with the past into the domain of irreality, leaving only the future light cone and the present event-point in the realm of reality.

past and yet affirm it of the spatially distant? From the relativistic point of view it seems clear that we are forced (not, of course, by logical consistency but only by plausibility of conceptual structure), if we are going to take an irrealist line as motivated by facts of epistemic distance at all, to certainly deny reality as fervently to the spatially distant as we do to past and future. What relativity does, with the invocation of spacetime to replace space through time, is not to force us in any way (Putnam, Rietdijk and others to the contrary) to reject our irrealist position, but only to symmetrize it in a natural way to include spatial separation on a par with temporal separation.

6

From the point of view we have been adopting, the doctrine of the irreality of past and future, taken as having a motivation in their epistemic distance, now seems clearer. We first reduce 'reality' to the lived experience of the observer; that is, we first fall, following a well-known verificationist path, into solipsism. Then seeing that our own future experiences and past experiences are as remote *now* from us as the spatially distant, the non-immediately sensed, etc., we fall from solipsism into solipsism of the present moment. Reality has now been reduced to a point!

Now obviously we don't want to go along this path. Where to block it is an interesting question. One can challenge the basic tenets of verificationism either with regard to its theory of knowledge or with regard to its semantic theory. One could certainly try to break the connection between the facts of epistemic distance, even if it is acknowledged to exist, and 'irreality'.

But what role does relativity theory play in all of this? One thing is certain. Acceptance of relativity cannot force one into the acceptance or rejection of any of the traditional metaphysical views about the reality of past and future. It can lead one to see more clearly than one did previously that by parity of reasoning one ought to treat spatial separation on a par with temporal. By forcing one in addition to say some things which seem peculiar in ordinary language it might lead one to move toward one position or another on grounds one didn't have before. But one who wishes to stick by an irrealist position, and is willing to pay the price, is certainly able to do so, all the while accepting the scientific reasonableness of Minkowski spacetime.

Much more interesting is the relationship between relativity and verificationism in general. Certainly the original arguments in favor of the relativistic viewpoint are rife with verificationist presuppositions about meaning, etc. And despite Einstein's later disavowal of the verificationist point of view, no one to my knowledge has provided an adequate account of the foundations of relativity which isn't verificationist in essence.

That one would want to do so seems fairly clear. Let me illustrate with just one problem. On the basis of verificationist principles one takes the attribution of distant simultaneity, of the isotropy of the velocity of light, etc. to be mere conventional decisions. On similar grounds, as we have seen, one can plausibly argue that the very adoption of the standard manifold topology for spacetime is merely conventional, any other topology saving the continuity structure along causal paths serving to 'save the phenomena' equally well.

But now consider the problem of so-called indistinguishable spacetimes. These are spacetimes which are counted as distinct by general relativity (they are by no means isometric to one another), yet which are such that no single observer can tell, even given a complete infinite lifetime, which of the spacetimes he inhabits. This could only be determined by a 'super-observer' who had access to the collected information of all the observers of the ordinary sort. Is it merely a matter of convention which of a number of indistinguishable spacetimes an observer inhabits? (Glymour 1972; 1977: Malament 1977b.)

Clearly, if one takes saving the phenomena as the sole task for which a theory is to be responsible, and a choice among theories which save the same phenomena to be a matter of mere convention, the answer depends on what one takes the phenomena to be saved to be. The pressure which drives us, verificationistically, to take the phenomena to be everything which is epistemically accessible to observers in general can also drive us to take them to be everything which is accessible only to one observer, oneself. This is, naturally, just the familiar slide from a phenomenalistic to a solipsistic position. Worse yet, should we not, given the epistemic inaccessibility *now* of past and future, take the phenomena to be rather the data of the one observer at the present moment, leading to a whole new range of indistinguishable spacetimes? Whereas before we counted as indistinguishable those spacetimes which appeared the same to any world-line in them, we now take them to be those which appear the same to single world-points. Surely verificationism has gone too far at this point.

But what I don't know is either how to formulate a coherent underpinning for relativity which isn't verificationist to begin with, or how, once begun, to find a natural stopping point for verificationist claims of under-determination and conventionality.

REFERENCES

Glymour, C. 1972. 'Topology, cosmology and convention.' *Synthèse* 24, 195–218.
Glymour, C. 1977. 'Indistinguishable space-times and the fundamental group.'
Foundations of Space-Time Theories, ed. J. Earman, C. Glymour and J. Stachel.
Minnesota Studies in Philosophy of Science, vol. VIII. Minneapolis: Minnesota
University Press.
Malament, D. 1977a. 'The class of continuous timelike curves determines the
topology of spacetime.' *Journal of Mathematical Physics* 18, 1399–404.
Malament, D. 1977b. 'Observationally indistinguishable space-times: comments
on Glymour's paper.' *Foundations of Space-Time Theories,* ed. J. Earman, C.
Glymour and J. Stachel. Minnesota Studies in Philosophy of Science, vol. VIII.
Minneapolis: Minnesota University Press.
Putnam, H. 1967. 'Time and physical geometry.' *Journal of Philosophy* 64, 240–7.
Rietdijk, C. 1966. 'A rigorous proof of determinism derived from the special theory
of relativity.' *Philosophy of Science* 33, 341–4.
Sklar, L. 1977. 'What might be right about the causal theory of time.' *Synthèse* 35,
155–71.

Modal reality

COLIN McGINN

This paper is given over to a difficult and controversial topic: how we should conceive the truth conditions of modal statements. Three interests motivate my concern with the nature of modal truth. One interest is semantical: what is the correct treatment of modal locutions in a meaning-theory for natural language? More specifically, what semantic role should a Tarskian theory of absolute truth assign to modal expressions? A second interest is metaphysical: what are the presuppositions and consequences of belief in objective necessity? Here I wish to include strict (or metaphysical) necessity and causal (or nomic) necessity, i.e. the alethic as opposed to the epistemic or the deontic modalities. I am interested, in particular, in uncovering the source of the uneasiness which a non-reductive view of such modalities provokes in philosophers of broadly empiricist persuasion. (The focus will be on so-called *de re* modality, but I think the underlying issue extends to *de dicto* modality.)

This second interest cannot fail to be intimately connected with the first, since how we choose to represent the truth conditions of modal sentences in a systematic semantical theory must reflect our considered conception of the nature of modal reality. That is, if we are to *assert* modal sentences, as is our common practice, then we thereby commit ourselves, metaphysically, to whatever our semantical theory attributes to such sentences as comprising their truth conditions. Conversely, our prior conception as to the constitution of modal reality should find its way into the semantics we propose, since modal sentences are the vehicle of expression for that conception. There is no divorcing semantics and metaphysics; we are obliged to take seriously whatever the favoured semantic theory pins on our serious modal assertions. In other words, instrumentalism about semantics implies instrumentalism with respect to the object-language. Insistence on this point has a significance beyond semantics for modal expressions, since an influential conception of the proper form of a meaning-theory for a whole language (model theory on possible worlds) is premised on a certain view of the semantic role of modal locutions: if that view proved philosophically suspect, then we might well have to abandon the derived conception of semantical theory.

The third interest I have in modal truth relates to the question of realism. On the one hand, I am interested in whether a uniform notion of realism can be formulated, applicable to all areas in which philosophical disputes have naturally been characterized in such terms; and modality is an area in which a significant modification of a certain standard formulation (that due to Michael Dummett (1978)) is needed if it is to have the required generality. On the other hand, there is an independently interesting issue about modal realism: how it is most naturally formulated, whether it is true, what view it requires of the relation between knowledge and reality. Coming at modality in this way will, I hope, illuminate the metaphysical questions surrounding essentialism and objective modality. The three interests converge on my opening question: how should we explicate that in virtue of which a modal statement is true?

My discussion has two main parts. The first part is addressed to what I shall call *objectual* interpretations of modal expressions: this is to be the thesis that the truth conditions of modal sentences introduce an ontology of possible worlds. It is often supposed that this just *is* realism about modality. I shall argue that, so formulated, modal realism is false; for there *are* no possible worlds in objective reality. The second part investigates *non-objectual* interpretations of modal expressions: this is to be the thesis that modal reality is not entitative in character, and the semantic role of modal expressions is accordingly not referential. Modal realism can still, however, be naturally formulated under a non-objectual interpretation, whereupon its threat to empiricist epistemology becomes manifest. So the first part of the paper deals with the location of modality in *ontology* – what objects there are; while the second part locates modality in *modes* – the manner in which objects have properties.[1] My guiding strategy will be to contrast modality with two other categories often thought illuminatingly analogous to it, viz. space and time. Use of these as a foil will help us get some perspective on the nature of modal reality.

I

According to possible worlds semantics for natural language modal expressions, the truth conditions of modal sentences involve commitment to a domain of objects (the worlds) over which modal expressions are taken to quantify. Formally, such ontological commitment can occur in two ways: either the vernacular modal sentences, containing (syntactic)

[1] This terminology is taken from Kit Fine's 'Postscript' to Prior and Fine 1977: 177.

modal connectives, are initially translated into a first order language in which the connectives are replaced by suitable restricted quantifiers and associated variables, so that a first order metalanguage can specify the truth conditions of the regimented object-language homophonically; or else the modal connectives are left primitive in the object-language and the possible worlds quantifiers occur only in the metalanguage, in which case the truth conditions are given non-homophonically. Such treatments have been proposed for the unary modal connectives and for the binary counter-factual conditional (see Kripke 1963; Lewis 1973). Under both sorts of formal treatment the sentences embedded within the scope of the modal connective are construed as *predicates* of the objects quantified over according to the stipulated truth conditions; so to make a modal claim is to make a (non-modal) predication on a domain of specifically modal objects. Thus the truth conditions of $\Box A$ are given by $\forall w(A'w)$ and $\Diamond A$ by $\exists w(A'w)$.[2] The modality expressed in the original sentence thus gets absorbed into a special range of values for a distinctive style of variable. Taken as part of a meaning-theory for a particular natural language, the suggested picture is then this: with respect to a class of expressions not overtly referential in form we are to discern an ontology of possible worlds, thus enlarging the domain of objects apparently spoken of, to which, in making modal assertions, the speakers attribute properties. This is, then, a semantical proposal in which we are invited to acknowledge the existence of entities whose inclusion in the ontology of our language did not immediately meet the eye: we become aware of the objects of modal discourse only when we set about contriving a theory that makes systematic sense of it.

In order to evaluate the ontological imputations of semantical proposals, we need some idea of the kinds of constraint to which such imputations are answerable: we need some regulative canons of ontological commitment. Discussion of possible worlds founders in the absence of agreed upon standards of ontological assessment. To get a handle on the issue I shall therefore recognize a number of conditions of acceptability which such ontological proposals may reasonably be required to meet. These conditions are not tendentious in the sense that their very formulation prohibits an ontology of possible worlds; rather, it will *turn out* that possible worlds have trouble meeting the conditions.

(i) *Formal.* By this I mean to cover a battery of requirements which ensure that the theory is semantically workable. Thus it is standardly required that a given semantical proposal allow a finitely axiomatized theory, that it do justice to the manifest structure of sentences, that it

---

[2] The primed sentence letter A' represents a possible worlds predicate corresponding to the sentence A; the notation is Fine's (Prior and Fine 1977: 119).

exhibit the language as learnable, that it account for structually valid inferences, that it satisfy some material adequacy condition such as Convention T, that it mesh with a model theory yielding a definition of validity, and so on. Meeting some such conditions as these is clearly necessary if the theory is to have any attraction at all.

(ii) *Linguistic*. This is a label for the following constraint: if a semantic theorist imputes a certain kind of ontology to a range of sentences not superficially committed to such an ontology, then he is under a *prima facie* obligation either to point to other expressions in the language explicitly so committed or to explain why it is that the ontology in question never surfaces. The motive behind this condition is to regulate and control departures from surface syntax. For it would seem implausible to discern reference to entities of a given kind in a speaker's language if he was *never* found invoking those entities explicitly. This condition is not perhaps absolutely binding, but failure to conform to it constitutes a strong presumption against the semantical proposal in question. On the other hand, if the condition is adequately met the theorist can adduce independent evidence for the correctness of his imputation; and certainly such corroboration can be used to justify a preference between formally equivalent semantic theories.

(iii) *Metaphysical*. The proposed ontology must not be intrinsically suspect; it must not comprise entities which are demonstrably metaphysically dubious. In particular, alleged values of variables should qualify as genuine individuals, on some acceptable explication of that indispensable notion. For to discern reference to entities of dubious metaphysical status would be to convict ourselves of error by our own lights. This might be formulated as a principle of *self*-charity: if sentences seem metaphysically innocent on their surface, then it is a count against a semantical theory of those sentences that it makes them metaphysically outrageous. A good example of a theory failing this condition is the Meinongian account of the semantic role of empty definite descriptions once favoured by Russell: it would be preferable to change our semantics rather than our metaphysics in this case, for we do not want to find ourselves speaking of entities which do not *exist*. I know of no conclusive argument that our language *could* not be thus aberrant, but I would prefer a theory that made my sincere and settled assertions true by my own standards.

(iv) *Epistemological*. The linguistic condition required that the expressions in question be related to other expressions of a certain syntactic type; the metaphysical condition required that the expressions relate acceptably to genuine extra-linguistic individuals; the epistemological condition imposes requirements on the relation between expressions and a speaker's *use* of them. That is, the attributed truth conditions should be pragmatically defensible, in the classical sense. More exactly, the introduced objects must play a suitable part in the learning and verification of the sentences concerned. The imputed ontology must dovetail with how the pragmatic phenomena observably are, and should not be alien to the phenomenology of understanding and using the sentences in question. In other words, what makes our sentences true should properly relate to our knowledge of their truth.

The foregoing conditions have been stated rather schematically. To see how they operate let us consider a semantical proposal that seems to pass the tests pretty well, and with which the objectual modal semantics unfavourably compares: this is Donald Davidson's (1967a) treatment of adverbially modified sentences. The basic idea of the proposal is that we see such sentences as containing implicit quantification over *events*; the adverbs are then taken as predicates of such events. Formally, the proposal is very like the possible worlds paraphrase of modal sentences. We cleave to first order regimentations by extending the domain of reference to include a range of objects not superficially spoken of; and just as predicates in modal sentences receive an extra argument place to accommodate the reference to worlds, so predicates in adverbial sentences acquire an extra place for the event variable. In both cases, then, a new style of bound variable is introduced and equipped with a new domain of objects to serve as values thereof. Now it is not my mission to defend Davidson's treatment of adverbs on all fronts, but I think it has some claim to meeting our four conditions. Here is a sketch of how it passes the tests.

Since I do not intend to confront possible worlds semantics on the score of formal adequacy, I shall not dwell on the formal merits (or demerits) of Davidson's theory of adverbs. Let it suffice to note that it seems to stand up quite well in respect of this condition. Davidson himself acknowledges the linguistic condition and argues, plausibly enough, that natural language is replete with referential constructions relating to events: quantifier phrases, definite descriptions, proper names, sortal predicates, pronouns, etc. And associated with these we have an elaborate apparatus of individuation: identity sentences, resources for counting, and so forth (Davidson 1969a). So imputing an ontology of events to sentences not obviously so committed can be corroborated by pointing to related areas of the language in which the ontology of events is explicitly and systematically invoked. Granted this we might explain the absence from the surface syntax of adverbial sentences of reference to events as the result of deletion transformations; whereas if corresponding referential constructions occurred nowhere in the language such an explanation would presumably not be available. Metaphysically, events seem to qualify as authentic individuals: they enter into causal relations; they can be given a clear identity condition; and most important, they admit of inequivalent descriptive identification, and so may be properly designated without being fully described, thus showing their independence from language.[3] Events seem as reputable as material objects as values of individual variables and subjects of predication: they are the kind of entity for which we can make sense of the idea of multiple property exemplification. We can also make sense of the distinction between particular and universal in application to events. Neither do they threaten trouble epistemologically. They figure straightforwardly in the explanation of how a speaker verifies sentences purportedly about events; indeed they are plausibly cited as *causes* of the perceptual beliefs which constitute (direct) verification of such sentences. They are thus directly implicated in the use of adverbial sentences, and are not such as to render knowledge of such sentences problematic. In other words, the suggested truth conditions for adverbial sentences afford a satisfactory link between truth and knowledge. So I think that Davidson's ontological imputation stands up well under our four conditions; indeed it departs little from the ontological paradigm set by material objects.

We must now consider how possible worlds compare with events in the above respects. I shall begin with David Lewis's account (1973: ch. 4), because it is the most forthright, clear and honest defence of possible worlds from a philosophical point of view that I know of. It should be remarked that, as Lewis notes, not much in the way of actual *argument* has ever been brought against the theory; and without such argument intuitive conviction must pale beside theoretical power. Arthur Prior vilified it as 'a tall story', but was reduced, in rejecting it, to relying upon ' "the choice of the soul" ', or, if you like, prejudice' (Prior and Fine 1977: 92–3).[4] I want to give the prejudice argumentative colour.

Lewis's view is that the quantified world variables which replace modal connectives range over a set of elements, homogeneous in themselves, one of which is distinguished as the actual world. Each such element *exists*; the actual world differs from the rest merely in its being the one we inhabit. The actual world is thus just one member of a more inclusive totality of objects not intrinsically different in kind. The word 'actually' is an indexical expression which, when uttered by inhabitants of other worlds, picks out the world in which it is uttered; actuality is thus a context-bound

---

[3] The idea that genuine individuals must be specifiable by inequivalent conditions is implicit in Davidson 1969a: 218.

[4] Prior took the same view of time: quantification over times is not to be interpreted literally either. As will emerge, I think there are strong reasons for an asymmetrical attitude towards ontologies of worlds and times.

property. As Lewis (1970) remarks, he intends this picture of modal reality to parallel a less controversial conception of the ontology of time: the present is just one element of a more inclusive totality of times each of which can be said (tenselessly) to exist.[5] The present is distinguished from other times simply in being the time at which we are – 'presently' is indexical. He might equally have compared his view of worlds with an even less controversial conception of space: the place at which I am is but one element of a larger totality of places, all of which (univocally) exist; but of course they do not all exist *here*, since 'here' is indexical. So Lewis wishes us to conceive of truth conditions for modal sentences strictly on the model of objectual truth conditions for temporal and spatial sentences: quantifiers over times and places function just as quantifiers over worlds. Keeping the comparison in mind enables us to take Lewis (as we must) absolutely literally.

Possible worlds semantics has undoubtedly enjoyed considerable formal success, largely because it allows techniques developed in classical extensional logic to be carried over to modal languages; and without its formal fecundity an ontology of possible worlds would (I suppose) hardly seem attractive. But formal success is not enough to establish its correctness; we need to be independently convinced that the introduced theoretical entities are acceptable, as Lewis himself recognises. So, first, do possible worlds meet the linguistic condition? Before I argue that they compare unfavourably with times and places in this respect, let me consider Lewis's own positive argument purporting to uncover direct evidence in natural language of (implicit) acceptance of his ontology of possible worlds. He writes:

I believe that there are possible worlds other than the one we happen to inhabit. If an argument is wanted, it is this. It is uncontroversially true that things might be otherwise than they are. I believe, and so do you, that things could have been different in countless ways. But what does this mean? Ordinary language permits the paraphrase: there are many ways things could have been besides the way they actually are. On the face of it, this sentence is an existential quantification. It says that there exist many entities of a certain description, to wit 'ways things could have been'. I believe that things could have been different in countless ways; I believe permissible paraphrases of what I believe; taking the paraphrase at its face value, I therefore believe in the existence of entities that might be called 'ways things could have been'. I prefer to call them 'possible worlds'. (1973: 84).

This argument proceeds by first finding an uncontentious modal sentence, observing that it contains a position open to existential quantification, and then construing the associated variable as ranging over possible worlds as Lewis characterizes them. Since Lewis is right that ordinary language allows such quantification, any fault in the argument

[5] The accompanying metaphysics of actuality is criticised in Adams 1974.

must come at the last step. In effect, Lewis is inviting us to regiment 'there are ways things could have been besides the way they actually are' by means of a first order quantifier somewhat as follows: '$(\exists x)(x$ is a possible world & $x \neq$ the actual world)'. However, this method of paraphrase seems to me neither obligatory nor natural. Consider the nonmodal sentence 'there are ways John is which annoy Jane'. This sentence would most naturally be taken to involve *second* order quantification into predicate position: for a way that something is is surely a property of it. Similarly for 'there is a way things are', where 'things' is some sort of device of plural reference; or alternatively 'there is a way the world is'. Now suppose we insert a modal operator into such second order sentences, as in 'there are ways John could have been besides the way he actually is'. This is naturally represented in the formula: '$(\exists F)(\Diamond F$ (John) & $\sim$ actually F (John)). This sentence says, not that there are *worlds* in which John (or some counterpart[6]) has different properties, but rather that there are *properties* that John might have had. Similarly, we can render Lewis's sentence thus: '$(\exists F)(\Diamond F$ (things) & $\sim$actually F (things))'. This says simply that there are properties that things could have possessed. In sum, what Lewis encouraged us to do with individual variables over worlds we can do (and do naturally) using second order quantification combined with modal operators. It woud be ineffective to protest against this alternative method of paraphrase on the ground that it employs primitive modal operators: for Lewis was trying to give an ordinary language argument *forcing* us to construe the sentences in question in his preferred way. So I deny that Lewis has drawn attention to a form of words that superficially requires, or even plausibly invites, the kind of truth conditions he wishes to propose. Certainly he has not produced linguistic evidence for his ontology comparable in richness to that available to the theorist of events.

Here we may observe a contrast between modality and space and time. For it is not difficult to discover a matrix of ontological locutions in which spatial and temporal expressions are embedded. Thus, in respect of space, we find locative indexicals, demonstratives, place names, functors forming complex singular terms for places, and quantification over places, standard and non-standard. And such ordinary language reference goes systematic in the advanced physical sciences. (Indeed, places have some

---

[6] Although Lewis substitutes a counterpart relation for the identity relation as between objects in distinct possible worlds (see especially Lewis 1968), this idiosyncracy seems logically independent of his position on the ontological status of possible worlds. However, it is very natural, once one has interpreted *de re* modal claims as introducing an ontology of counterparts, to take modal predications on the actual world as similarly introducing numerically distinct counterpart worlds. I find the latter move just as implausible as the former, but I will not argue against Lewis on this ground.

claim to be ontologically more basic than material objects in our language.) In the case of time, the language seems similarly entitative: associated with the tense modifiers we have a systematic nomenclature for times (the system of dating), as well as temporal demonstratives, functors, quantifiers, etc. Accordingly, a semantics for spatial and temporal expressions employing quantification over places and times can be corroborated, like the ontology of events, by observing the manifest presence of such ontological commitment in the object-language. But in respect of the possible worlds ontology modal language seems pointedly impoverished: no proper names for worlds, no functors, demonstratives, and no clear cases of quantification (standard or non-standard) over the alleged worlds. Nor do we find ready talk of identity and number in application to possible worlds. Yet surely if there are such entities and we daily quantify over them, our language might reasonably be expected to extend to overt recognition of them; at the very least, there is a mystery here for Lewis to dispel.

The availability of ontological locutions for space and time seems predicated on their dimensionality. The three spatial dimensions permit a coordinate system, crude or sophisticated, with respect to which places may be identified; spatial functors trade upon this. Similarly, our system of dating exploits the linear ordering of time. That is to say, for places and times we have the idea of an ordered domain of elements, where the kind of ordering is reflected in our means of reference to those elements. In the case of worlds such an ordering idea seems arbitrary at best; so it is hard to see how a system of reference, comparable to that applicable to space and time, could be successfully and usefully introduced. The more inclusive totality of which Lewis speaks does not present itself as an arrangement of objects along some natural dimension (or dimensions) allowing for identification according to position along that dimension (those dimensions) relatively to other elements.[7] So not only is there *in fact* no clear referential apparatus for possible worlds; it is hard to see how there *could* be, at least after the fashion of space and time. And this starts a suspicion that the ontology of worlds is alien to the conceptual resources of the ordinary speaker; it is not an ontology for which we can be said to harbour an antecedent predilection. In fact, the modal connectives seem, in the present respect, closer semantically to the truth-functional connectives.

If what I have just said is right, we do not pretheoretically recognize possible worlds as genuine individuals; but is there direct reason to withhold this appellation from them? If there is, then possible worlds are

[7] Lewis himself notes the contrast between worlds and times in respect of ordering in 1973: 105.

ineligible as values of individual variables and subjects of predication; so the first order regimentation of modal sentences would inaccurately represent the modal facts. Now it is not easy to *define* the notion of an individual, vital as that notion is, but the following two conditions seem necessary to individuality: something is a genuine individual only if (a) it admits of proper identification short of exhaustive characterization, and (b) its properties partition (non-trivially) into the essential and the accidental. Condition (a) captures the idea that an individual is an extra-linguistic entity whose properties exceed those we happen to fix upon in referring to it: and this is essential if we are to apply the picture of first identifying an individual as a potential subject of predication, and then informatively characterizing it by coupling the identifying singular term with a predicative expression. Condition (b) tells us that an individual is something that has certain properties essentially, but is also such that it can exist through variation in respect of other of its properties. Both conditions ensure that an individual is something distinct from the descriptions true of it. (Meinongian objects seem to fail these conditions.)

Now I think we have enough of a grip on the notion of an individual to appreciate that possible worlds are unqualified for that status; whereas places and times on the other hand adequately meet the conditions. Kit Fine (Prior and Fine 1977:158) remarks of times that they enjoy scarcely any essential properties but these: that they are times, that they exist, and that they are ordered in a certain way – the rest is accidental. It follows that it would not be feasible to identify a time with the conditions which obtain at that time, i.e. with what is true at it. A parallel observation applies to places: only their existence and their positional properties seem essential to them – what goes on at a place is not constitutive of its identity. Moreover, times and places may be picked out in inequivalent ways, and uniquely identified without complete description. So times and places may be made to figure as genuine subjects of informative predication. But with worlds none of this is so. For as Fine (158) goes on to note (though not to make the present point) what transpires in a world *is* essential to its identity; the identity of a world is fixed by its 'content'. This implies that worlds do not permit a (non-trivial) partition of their properties into essential and accidental; so one loses one's grip on the idea that a particular world is distinct from the set of properties which characterize it.

Furthermore, it seems clear that a world has not been *uniquely* specified until all of its properties have been listed; it is not determinate of which world one is speaking until its whole content has been specified. It is difficult to see, therefore, how we can apply the idea of first picking out a world and then predicating some property of it. (Another way of putting the point is that we cannot easily make sense of informative identity statements about worlds, since it seems that they can be presented to us in just one way.) We can perhaps conceive of the *actual* world as an individual, but that is precisely because we think that *it* could have been different. Indeed, it seems more natural to construe what are called possible worlds as ontologically of the nature of states or properties; but if so, modality belongs rather with predicate position: it is not properly associated with values of individual variables. (Space and time seem quite otherwise.) So formulas containing alleged world variables should be viewed with suspicion: we do not understand their import just because we can write them down – we must satisfy ourselves that they can really mean what they purport to. It seems to me that the indicated problematic status of possible worlds as individuals renders unsurprising the observed reluctance of our language to treat them so. (I shall return to this question in another connection.)

The fourth condition we imposed on ontological imputations was epistemological. And here I think that the Lewisian truth conditions encounter serious trouble: for they yield a conspicuously incorrect epistemology of modality. Before I give my reasons for saying this, let me distinguish the objection I want to urge from a different epistemological worry one sometimes hears voiced. The worry is this: possible worlds are entities causally insulated from each other, in particular from how things are in the actual world; but then how can we come to know about them, since knowledge about a range of objects requires causal interaction with those objects? In other words: we have some modal knowledge, but this could not have been acquired by inspection of possible worlds, owing to their causal insulation; so possible worlds are not implicated in the epistemology of modality; therefore the truth conditions of modal sentences do not concern possible worlds. The general point here is that we cannot get into epistemic contact with entities so remote from the sphere of actuality in which we are condemned to toil.

Now I would be the first to agree that there is a genuine perplexity behind this worry, but I doubt its dialectical power against a proponent of possible worlds. There are two points. First, as I shall notice below, modality gives rise to such epistemological problems even when non-objectually construed; so the underlying difficulty is not escaped by abolishing the worlds. Second, there are plausible semantical theories of other types of sentence which present exactly analogous perplexities – the most notable being mathematical sentences. If the truth conditions of mathematical sentences involve abstract entities, themselves causally inert, then a parallel epistemological problem arises.[8] But since the

[8] See Benacerraf 1973 for a clear statement of the issue.

alleged problem is not *distinctively* presented by possible worlds semantics, a defender of those entities can produce some motivation for resisting the conception of knowledge that the objection rests upon. At least, it could be maintained that we have to do with a genuine antinomy here between our metaphysics and our epistemology, and not a knockdown objection to that metaphysics. So if we are to dispatch possible worlds on the score of epistemology, we need considerations of a more specific and less question-begging sort.

Let us grant, then, as a concession to Lewis, that we possess some cognitive faculty which enables us to know what is true in other possible worlds; a faculty of the same species as that proposed by some (e.g. Gödel) in respect of mathematical entities and structures.[9] To fix ideas, we might think of the faculty as a kind of mental vision, analogous perhaps to the ordinary vision whereby we learn of the actual world. Now, as I said, Lewis takes modal sentences to contain quantification over a homogeneous domain of objects, the worlds. So the question for Lewis is how the postulated modal faculty relates to the entities comprising the truth conditions of modal sentences so as to yield modal knowledge. According to Lewis (1973:90), the cardinality of the set of possible worlds is equal to or greater than the cardinality of the set of subsets of real numbers: so the quantification is over a non-denumerably infinite set of objects, the same in kind, remember, as the actual world. Now let us enquire how sentences with such quantificational truth conditions are verified.

Our basic model of the verification of such sentences is, as Dummett says (1973:236–9, 634–6), the 'direct method' of applying to each element of the domain some test which determines whether or not it satisfies the condition in question. (If the condition embeds further quantifiers, as would a possible worlds paraphrase for iterated modalities, this may involve more than a single such total check.) It seems plain, however, that we *could* not verify modal sentences by the direct method, since the modal faculty would presumably require a non-zero time to test whether a given world satisfies the condition appended to the possible worlds quantifier. (Just think how long it takes to verify a necessary empirical truth in the actual world!) So our knowledge of the truth of a sentence of the form $\Box A$ could not be acquired by directing the modal faculty sequentially and exhaustively to the infinite domain of possible worlds. If that were the only available method of verification, then modal claims would (or should) be the object of extreme scepticism, since our evidence for modal statements would perforce fall far short of covering that in virtue of which

---

[9] For a discussion of the Gödelian faculty of mathematical intuition, see Steiner 1975: 130ff.

they are true. My point here is not, of course, that no sentence can be credited with such radically undecidable truth conditions: it is rather that, since modal sentences manifestly do not thus transcend our powers of verification, an account of their epistemology on the present lines cannot be correct. Modality contrasts strikingly with space and time in this respect: contingent sentences about all places and all times typically *do* present severe problems of decidability. So the quantificational truth conditions of such sentences well match the extent of our spatial and temporal knowledge: we find just the degree of undecidability by direct method that the suggested truth conditions predict.

The possible worlds theorist must, therefore, if he is to account for our modal knowledge, suggest some *indirect* means by the exercise of which our modal knowledge might be acquired and justified, where this means suffices to yield knowledge of what holds true in the infinity of possible worlds. I do not know of any actual suggestions to this end, but those that I have devised seem to me not to work. What is needed is some method having two components: a principle for selecting some subset of worlds which the modal faculty can be reasonably supposed to survey; and a rule of inference which licenses us to move from direct knowledge of the properties of that subset to justified conviction concerning the containing totality. A straightforward suggestion, then, is the method of ordinary enumerative induction: we proceed by surveying mentally some finite number of selected worlds with a view to determining whether the sentence in question is true at those worlds; and if it is, we infer that the remaining unsurveyed worlds follow suit. Thus it is that we may come to know that some statement is necessarily true. Such a method certainly seems phenomenologically inaccurate; but it is also inadequate, as a matter of principle, to account for our modal knowledge. For surely the subset of worlds we can reasonably be supposed to check in the time it takes to arrive at a justified modal belief will fall far short of exhausting the set of worlds in which the sentence is required to be true. If this were the procedure, then our grounds for modal claims would be highly non-conclusive: we would have no right to the confidence we commonly repose in such claims, and scepticism would be indicated. (Recall that many necessary truths have been supposed paradigms of certainty.) So this suggestion misrepresents the epistemology of modal truth. Again, it is instructive to contrast modality with space and time: applying such an inductive method to a large totality of places and times leaves us, and rightly so, in a condition of extreme dubiety.

Our degree of conviction about modal truth might now prompt the idea that our method of establishing it might resemble that of inductive proof in mathematics; for here we attain a comparable confidence about comparable domains. Such a method would require a basis premiss, perhaps establishing truth in the actual world, conjoined with an induction step, exhibiting the truth of a sentence at the $k + 1$ world as dependent upon the truth of that sentence at the kth world, thus demonstrating the truth of the sentence at all worlds. This method at least seems to be the right *kind* of justification procedure to explain our confident modal knowledge, but it does not appear feasible in the present case: for the worlds are not well ordered under any satisfactory ordering relation, as the natural numbers are ordered under the successor relation, and I can see no plausible way in which the induction step might be proved. In default of a detailed and specific proposal of this sort, I therefore conclude that this method will not bring the totality of worlds within the reach of our circumscribed epistemic capacities.

Seeing that inductive methods are unworkable, someone might suggest a falsificationist procedure for the discovery of modal truth. Thus we do better to direct the modal faculty upon the subset of worlds most likely to falsify the given statement: if the statement passes this test, then our rule of inference licenses us to accept the modal sentence as true. The analogy with space and time is that we go to the place or time at which the universal sentence will most likely be falsified, there make the requisite observations, and judge accordingly. It is not altogether clear how we might set about locating such a potentially falsifying world, but anyway the suggestion has the same defect as the inductivist proposal: our modal beliefs should be relegated to the extremely tentative, as tentative as we are often told our scientific beliefs should be; but I think it is obvious that no such insecurity is right. The general difficulty is that the gap between the inspectable subset of worlds and the total set does not seem bridgeable by some principle of inference which justifies our actual modal beliefs. It therefore seems to me that either the epistemology of modality cannot conform to the possible worlds truth conditions and so those truth conditions must be rejected, or it should be said (heroically) that our modal convictions are unjustified and should hereafter become victims of scepticism. Since I have argued against possible worlds on other grounds, and since I think there is an alternative view of modal truth, I advise that we retain our modal beliefs and extrude the worlds.

It is natural now to suggest, if one is impressed by the preceding difficulties, that the epistemological condition be met by trying to divorce truth conditions and verification conditions. The trouble arose because we took the verifying modal faculty to operate by apprehending the objects whose properties determine modal sentences as true or false. The solution, it may be suggested, is to dissociate the modal faculty from those objects, and let the grounds of modal knowledge consist in something other than (direct) knowledge of the worlds. The suggestion is the analogue of a certain proposed solution to the problem of mathematical knowledge: since mathematical entities construed platonistically seem unknowable on a causal theory of knowledge, we must ground such knowledge, not on any direct apprehension of their properties and relations, but on some empirical basis. Thus it is suggested that we know mathematical truth by perception of proofs, these understood as stretches of notation; or again, that the application of mathematics in empirical science provides a route to mathematical knowledge, since we verify our empirical theories holistically.[10]

My general complaint about such suggestions is that, in severing the connection between what renders the sentences in question true and the means by which we come to know their truth, we make a mystery of how it is that our having just those grounds can constitute a proper justification for knowledge of truths of just that sort. In other words, it is hard to see how a person can justifiably believe a sentence if his justification is unrelated to that it is offered as a justification *for*: for what makes it the case that he justifiably believes a sentence with *those* truth conditions?[11] Certainly, as a matter of general principle about knowledge we take it that some such connectedness condition has to be met. But in case this general point fails to carry conviction, let me examine a specific and tempting proposal along these lines for the case of modal knowledge.

It is very plausible that, at least for the strict modalities, knowledge of the modality of a given sentence is arrived at *a priori*. This is pretty evident for sentences whose truth (as distinct from their necessity) is known *a priori*, but it also seems to hold for necessary *a posteriori* sentences, e.g. statements of natural kind, composition, identity and origin. Thus it may be said that we do not gain knowledge of the corresponding modalized sentences by quasi-empirical inspection of the possible worlds in virtue of which the sentences are true: rather we come to know them by some sort of grasp of concepts. More precisely, we come to know that a certain empirical statement is necessary by inference from a pair of premises: the first is the non-modal empirical truth which we know by ordinary *a posteriori* procedures; the second is a conditional, affirming that if the concept in question applies to a sequence of objects then it does so necessarily, where this conditional is known *a priori* by reflection on the

---

[10] For such a view of mathematical knowledge see Hart 1979: Hart's position derives from Quine's suggestions in 'Two Dogmas of Empiricism' (Quine 1961: 20–46).

[11] This kind of point is urged by Benacerraf 1973: 672–3, against epistemology for (platonistic) mathematics based upon the empirical accessibility of proofs; I am urging the point more generally.

concept in question. *Modus ponens* delivers the modal conclusion.[12] Now the suggestion is to be that this account of such modal knowledge nowhere cites singular premisses about possible worlds, i.e. instances of the universal quantifier to which the modal connective is equivalent: instead it speaks of concepts and our grasp of them – so the modal faculty relates not to worlds but to concepts.

Now it is not that I think this account of modal knowledge is wrong – though it is often mischaracterized – but I do not see that it helps the possible worlds theorist. Note to begin with that this entirely rescinds the analogy with space and time, so far as the epistemology of the statements is concerned. Secondly, the second premiss is itself modal, so knowledge of it is knowledge about all worlds: we must then say that *it* is known without adverting to how it stands in those worlds. But if so my general difficulty becomes acute: for how could 'grasp of concepts' warrant belief about the infinite set of worlds? The concepts onto which the modal faculty is directed presumably exist in the actual world; so it is hard to see how such grasp could inform one of how things are in all the possible worlds. That is to say, the source of modal knowledge is being located in a region of reality which is both quite distinct from, and inexplicably related to, the region whose condition the modal knowledge is knowledge of. It is therefore quite mysterious how we can justifiably pass from the alleged grounds to that for which they are grounds. I suppose that someone strongly attached to possible worlds could choose to live with this mystery: but it seems to me to be a powerful *prima facie* objection to the possible worlds truth conditions. One would prefer, other things being equal, to have a theory of modal truth conditions which did not have this unattractive consequence; such a theory is proposed in the second part of the paper.

I have argued that Lewis-style truth conditions for modal sentences fail a number of well-motivated conditions on the acceptability of ontological imputations. In doing so I have had to spell out views and contrive possible defences to the objections I have raised. This runs the risk of incurring accusations of *ignoratio elenchi,* or of encouraging retreat to positions seemingly less vulnerable. My own suspicion is that Lewis's metaphysics is the only way to make clear and honest sense of an ontology of possible worlds, but I am aware that this will not meet with instant agreement. So I should consider some alternative views of possible worlds which have been supposed to avoid the excesses of Lewis but which preserve the intuitive content of the objectual view. I shall therefore

---

[12] This is the account of our knowledge of *a posteriori* necessities naturally suggested by Kripke's (1972) discussion of essentialism.

discuss the conceptions of possible worlds suggested by Robert Stalnaker (1976) and by Saul Kripke (1972: 264ff.).

Stalnaker's theory is developed in reaction to Lewis's. Stalnaker wants a conception of possible worlds which recognizes their literal existence as irreducible entities, and which serves in the interpretation of modal locutions and in the analysis of intensional concepts such as that of the proposition. He wants, that is, to preserve the idea of possible worlds as values of individual variables. But he rejects Lewis's picture of a set of worlds not differing in intrinsic nature from the actual world. (We might put it by saying that he rejects the analogy with space and time.) Stalnaker agrees with Lewis that worlds are 'ways things might have been', but denies that these entities are of the same sort as the actual world. In fact, he seems to hold that possible worlds are to be conceived as states *of* the world. So the world variables are taken to range over such possible states. Now I have already argued that the form of words invoked by Lewis (and now by Stalnaker) should not be interpreted as containing a first order quantifier over possible entities, but should instead be seen as a second order quantifier over non-modal properties which are predicated in some modality of ordinary actual objects. But let us for the moment go along with the alternative regimentation: then we can ask whether the attendant ontology of *states* is acceptable.

I have three sorts of objection. First, this proposal does not seem to escape two of our earlier complaints: we are still short of convincing linguistic corroboration of the introduced ontology; and the epistemological problems seem to recur in precisely analogous form, since the verification of modal sentences will now involve surveying the infinity of possible *states* or employ some clearly inadequate indirect method. Second, the metaphysical objection from what it takes to be an individual is avoided only by undermining the title of Stalnaker's worlds to feature as values of individual variables. If his possible worlds are of the nature of states or properties, then they belong with expressions whose semantic role is that of predication; but to concede that is to give up the quantificational paraphrase. There is a related query: it is customary to view truth in the actual world (truth *simpliciter*) as following by universal instantiation from truth in all worlds (necessary truth); but this seems possible on Stalnaker's view only if we distinguish the actual world from the world. If possible worlds are states and the world is an individual, then the actual world must itself be a state of the world if it and the other worlds are to comprise a homogeneous domain over which universal instantiation can be interpreted as valid. But surely part of the appeal of Stalnaker's picture was that the actual world is distinguished from merely possible worlds just by being *the world* – 'I and all my surroundings', in

Lewis's phrase. Thirdly, Stalnaker advertises possible worlds, as he construes them, as entities capable of giving a non-circular explanation of propositions. If worlds are taken in Lewis's way, one can appreciate the motive: a proposition becomes a set of concrete (though non-actual) individuals. But on Stalnaker's explication the appearance of explanation seems to lapse: propositions become, in effect, sets of properties. But properties are universals, the kind of entity of which propositions are said to be composed, so the circle of explanation begins to look small indeed. In fact, given a nominalist view of universals, Stalnaker's conception of possible worlds seems to collapse into a sets of sentences view. It would be odd if Stalnaker's 'moderate realism' about possible worlds were to depend upon rejection of such nominalism.

I therefore think that the Stalnaker view is more eliminative or reductive than he intended it to be. The view, indeed, often seems to amount to little more than the anodyne suggestion, acceptable to those who refuse the objectual truth conditions, that the world might have been different; which is just to say that the world is an individual with modal properties. That is harmless enough, but in no way warrants quantification over an infinite domain of possible worlds. It seems to me no better than inferring from the fact that a person might have had a different life history that there are infinitely many possible persons corresponding to the actual one who *are* enjoying those alternative life histories, either as individuals or as states. It is less easy to find a halfway house between primitive modal connectives and Lewis's 'extreme realism' than one might suppose. At any rate, Stalnaker's suggestions fail to justify the objectual truth conditions he aimed to defend.

Kripke is popularly supposed (a) to have scouted a Lewisian conception of possible worlds, (b) to have offered a picture of the ontological status of possible worlds which is metaphysically acceptable, and (c) to have supplied, by way of (b), a satisfactory objectual interpretation of the truth conditions of modal sentences. In other words, Kripke is taken to have given a metaphysically innocent account of the domain of objects over which possible worlds variables range, where the truth value of a modal sentence turns upon conditions of those objects. I think this popular view of Kripke's views is mistaken. However, the conflicting pressures at work in Kripke's discussion make it difficult to extract a coherent doctrine: he can be found consistent, but only by modifying the popular (and somewhat natural) view. Indeed, his ultimate position, it seems to me, is not calculated to promote the cause of possible worlds semantics.

On the one hand, he is anxious to deny that possible worlds are correctly conceived as 'other dimensions of a more inclusive universe'

(1972: 345): this amounts, fairly clearly, to a rejection of the standard parallel with space and time, since the here and now clearly *are* elements of larger totalities. He prefers, instead, to speak of possible worlds as *postulated* entities, not as objects of intellectual *discovery* at all (1972: 267).[13] Possible worlds are claimed to be, in Lewis's phrase, 'creatures of the imagination'. This conception is directly opposed to Lewis's 'realist' position, since possible worlds are not taken by Kripke to exist independently of our imaginative stipulations. It strongly suggests an analogy with the intuitionists' conception of the ontological status of numbers, viz. that they are 'mental constructions'. If so, Kripke's characterization of possible worlds implies that quantified world variables range over a domain of *constructed* objects: and on that intuitive interpretation of modal sentences we would expect the standard consequences of such anti-realist views – namely, intuitionistic-type logic and time-relative truth.[14] In fact, the Kripkean picture of possible worlds can be seen as a reduction of the truth conditions of possible worlds sentences to their assertibility conditions, since exercises of imagination constitute (for Kripke) our route to knowledge of possible worlds. These observations have the following consequence: *if* the truth conditions of modal sentences consist in facts about possible worlds, then on Kripke's conception of the latter, modal truth should be taken anti-realistically.

That is indeed a consequence that many would happily embrace: but not Kripke. For, on the other hand, Kripke clearly wishes to maintain a strongly objectivist view of modal truth: his essentialism commits him to it, and he often speaks, realistically, of our 'seeing' whether some statement is necessary or contingent (e.g. 1972: 267). That we are at liberty to *stipulate* modal truth would be anathema to him; rather, we must discover what is necessary by philosophical reflection ('intuition') and scientific enquiry. The tension is obvious: how can possible worlds comprise the truth conditions of modal sentences if the former are anti-realistically conceived while the latter are interpreted in a realist way?

It seems to me that consistency requires that Kripke deny the correctness of such truth conditions: that is to say, modal sentences should not be given a possible worlds semantics, assuming that the job of semantics is to specify the conditions of strict and literal truth. What is interesting is that Kripke does come close to concluding as much. After lamenting the abuses to which the notion of possible world has been put, he says: 'It is better still, to avoid confusion, not to say "In some possible world

---

[13] Rescher (1975) elaborates what he takes to be a similar view.
[14] Rescher remarks upon this seeming consequence (1975: 97).

Humphrey would have won [wins]" but rather, simply, "Humphrey might have won" ' (1972: 345). This very strongly suggests that, if we are limning the true and ultimate structure of modal reality, we do better to leave the modal connectives primitive: paraphrasing them with possible worlds quantifiers misrepresents how things modally are. Prior took a similar view: if we are to talk of possible worlds at all we should construe such talk as strictly derivative from talk in which modal expressions are semantically non-objectual, for the primitive modal connectives better convey 'the structure of the facts'.[15] So it seems that if Kripke is to bring his modal semantics into line with his metaphysics he needs to favour a position like Prior's, in which case a semantics must be provided which really conforms to the advice Kripke offers in the remark quoted above.

In short, Kripke is not in fact a modal objectualist: the truth conditions of modal sentences are determined prior to, and independently of, the imaginative construction of possible worlds. What, then, is the status of possible worlds for Kripke, both in his informal remarks and in his formal model theory for modal logic? I think the only answer can be that we are to take such talk merely as evocative metaphor: its significance is purely heuristic, in the sense that it can aid our thought about matters modal. But it does not reflect the sober metaphysical truth. This means that all such talk must be ultimately eliminable when employed in serious contexts, on pain of talking fiction.[16] So possible worlds model theory cannot be understood as providing a genuine *interpretation* of serious modal discourse; what it gives is just an algebraic formal model. That is what I meant when I said that Kripke's view of possible worlds undermines the ambitions of possible worlds *semantics*. Perhaps possible worlds, understood as imaginative mental constructions, have a place in an account of how we *verify* modal sentences, but they cannot, so understood, supply a foundation for a realist conception of modal *truth*. It seems to me, therefore, that Kripke's view of the ontological status of possible worlds is basically right, but that the view disqualifies them from featuring in

---

[15] The phrase is Prior's (Prior and Fine 1977: 54). (I should perhaps make it clear that my informal talk of *facts* in this paper is not intended to imply any commitment to an irreducible ontology of such entities.)

[16] One notable context in which Kripke permits himself to quantify over possible worlds is in his characterization of rigid designation – a is a rigid designator of x iff a designates x in every possible world in which x exists. However, such quantification seems easily eliminable in favour of modal operators, thus: a is a rigid designator of x in L iff a designates x in L & $\sim\Diamond\,\exists\,y\,(y \neq x$ & a designates y in L). (This definition can be employed for rigidity with respect to modal operators weaker than, or at least different from, metaphysical modalities: thus we have *nomically* rigid designators, e.g. 'the gravitational constant', 'the speed of light', etc.) On the other hand, terms naturally described as temporally rigid or even spatially rigid may, I think without metaphysical impropriety, be defined by quantification over times and places. There seems nothing objectionable about defining these different kinds of rigidity in formally different ways.

modal truth conditions; so I need not rehearse my earlier objections to possible worlds in application to Kripke's conception.

I have now reviewed what I take to be the principal candidates for a philosophical explication of the ideas on which objectual modal semantics is based, and found them wanting. What are the consequences for realism about modality? We can say at once that if modal realism depended upon construing the subject matter of modal sentences in terms of possible worlds, then indeed modal realism would be unacceptable. However, I think an alternative formulation of modal realism is available which does not encounter the difficulties of the objectual formulation; I turn to this in the second part of the paper. But a second point has emerged: possible worlds semantics is not *sufficient* for modal realism either, since quantifiers over them might be interpreted in the manner of the intuitionists. Just as the intuitionists employ (non-substitutional) quantifiers over numbers but construe them constructively – numbers are the products of creative mental acts – so an adherent of possible worlds semantics might propose a constructivist interpretation of his quantificational metalanguage and assume the underlying logic to be intuitionistic.[17] The values of world variables would then comprise something like imaginative acts or their contents, in which case the truth of a modal statement would consist in the obtaining of the conditions in which we *recognize* such a truth. We must, accordingly, think again about the truth conditions of modal sentences and about what it is to hold a realist view of modal truth.

2

The quantificational treatment of modal expressions naturally corresponds to the idea that the metaphysical category of modality is that of *object*. I said that the opposing view conceived modality as consisting precisely in *modes*: but what is a mode? The intuitive content of this notion is that of a *way* of possessing properties. I should like to put this as follows: modalities are to be conceived as higher order conditions on properties – they are properties of properties. Thus we can say that an object has a property dispositionally, or that a property is an essential property of objects which instantiate it, or that the properties expressed in a law of nature are related by nomic necessity, or that an analytic sentence has the

[17] Given the constructivist conception of possible worlds I have attributed to Kripke, and given his 'possible worlds semantics' for modal languages, one might think it natural for Kripke to integrate his semantics for modality with his semantical interpretation of intuitionistic logic, as developed in his (1965) 'Semantical Analysis of Intuitionist Logic I'. How this would work out I leave to others more competent than I to judge: but clearly it would imply an anti-realist view of modal truth, if taken as an account of literal truth conditions.

property of truth necessarily. So understood, modal qualifications can be compared with other locutions naturally construed as conditions on properties. Quantification itself, as Frege characterized it, functions as a higher order property (second-level concept). (But this *analogy* between modes and quantification should not be construed as actual sub-sumption – modality is not a *kind* of quantification.) Value qualifications seem to behave analogously too: we say that it is good to have a certain property or bad that a pair of properties be combined. (So a comparison with deontic expressions can be made independently of possible worlds truth conditions for the two sorts of locution.) Or again, modality might be compared with the properties of objectivity and subjectivity: these notions are best seen as properties of properties – they categorize first order properties of individuals.[18] It seems to me that this conception of the metaphysical character of modality captures the intuitive idea of a mode of property instantiation; and it is quite removed from an objectual view of the import of modal expressions. Moreover, it well consorts with the syncategorematic character of modal expressions: they qualify other predicates, while not themselves capable of ordinary first order predication (save by ellipsis). I am concerned, then, with the question of realism with respect to facts of this higher order structure (if I may so speak). But before I address the matter directly, some remarks are in order on semantics proper.

I take it that one constraint upon semantic proposals is that, in a sense it is difficult to make precise and hygenic, they should do justice to our prior substantive conception of the sector of reality with which the sentences at issue are designed to deal. So we need a semantics for modal expressions which conforms to the picture of modal facts just sketched. I am myself somewhat doubtful that such a semantics is presently available, but I think there are formally workable proposals which are at least in the right spirit. I shall mention three kinds of theory; this will at least show that the alternative to possible worlds semantics need not be a refusal to theorize, as Lewis (1973: 85) insinuates.[19] Each proposal can be said to assign a distinctive semantic role (contribution to truth conditions) to modal expressions, though they treat them as primitive in the sense that their semantic content is not in any way *analysed*.[20] (i) One style of truth

[18] Here I have in mind Thomas Nagel's (1979) discussions of subjective and objective: see 'What is it like to be a Bat?', and 'Subjective and Objective'. It is a notable fact that all four sorts of higher order property have been objects of suspicion or of reduction: perhaps there is some general anti-realist tendency to doubt the reality of facts of this apparent structure.

[19] Would he count the standard Tarskian clauses for the truth-functional connectives or for the classical quantifiers as instances of theoretical abstinence?

[20] I am assuming the Davidsonian distinction between attributions of logical form and conceptual analyses of structurally primitive semantic elements: see, e.g. Davidson 1967b.

theory regards modal expressions as genuine operators on open and closed sentences by employing an intensional metalanguage in which modal operators modify sentences containing semantic machinery. Formally the method parallels the usual Tarskian clause for negation, and in a clear sense yields homophonic truth conditions (e.g. Baldwin 1975; Peacocke 1978; Gupta 1978; Davies 1978). (ii) There are recommendations to see modal expressions as more properly attaching to *predicates* (or predicate abstracts): this gives natural expression to the idea of a modal property, and may avoid some alleged problems about permitting bound variables in the scope of modal operators. This approach recognizes a category of predicate modifiers and enriches the usual truth-theoretic resources to handle these.[21] (iii) One might extend Davidson's (1969b) paratactic theory of propositional attitude expressions to modality.[22] This approach keeps the theory extensional, and again sidesteps issues about opacity. It would need to accommodate quantification into modal contexts, of course, but any solution for the case of propositional attitudes would presumably go over to modality. On such a treatment, modal expressions would be metalinguistic. A related metalinguistic view, due to Quine (1979), treats modal expressions as multigrade predicates, satisfied by objects and predicates. None of these theories requires quantification over specifically modal objects; and each seems at least consistent with the metaphysical picture I favour. Since the matter is controversial, and since my present concerns do not call for a firm decision, I shall not try to defend a particular theory here; but for the sake of definiteness the reader may select theory (i) as the semantic background to what follows.

Realism with respect to a given class of sentences has been characterized as the thesis that the truth conditions of those sentences transcend the recognitional capacities possessed by their users.[23] Let me distinguish three kinds of recognition transcendence. First, the class of sentences in question might be such as to permit us evidence bearing upon the truth value of any sentence of the class – so that we can always put ourselves in a position justifiably to assert or deny any given sentence – but the evidence may always fail to be conclusive: that is, the class admits of complete but non-conclusive verification. Second, we cannot guarantee

---

[21] A theory of this type is developed in Peacocke 1976.

[22] I would like to extend the paratactic theory to the relational case by letting the quantifier in the first sentence actually *bind* the variable free in the second. This suggestion is encouraged by cases of pronominal cross-reference like 'There is a big fish in the river. Go and catch it!', which are also not plausibly construed as truth-functional compounds of open and closed sentences. But I cannot elaborate on the suggestion now.

[23] This is Dummett's (1978) general formulation of realism. He takes this to result from insistence on bivalence for undecidable sentences: however, it should be noted that recognition transcendence could obtain in cases which admit of complete though nonconclusive decidability.

complete verifiability – some sentences of the class may altogether elude evidentially warranted assertion – though *some* sentences do permit recognition of their truth or falsity. Third, the sentences are *always* unverifiable – even non-conclusively – by means of our actual recognitional faculties. Each of these possibilities may be described as cases in which the realist view of truth conditions introduces a gap between the world and our knowledge of it. The anti-realist characteristically protests at the introduction of such a gap: for, on a realist view, the relation between knowledge and reality becomes problematic. So we can say that a realist suggests a conception of reality to which our epistemic faculties may be in some way inadequate, whereas an anti-realist conceives the world as *essentially* accessible to our epistemic faculties.

Now what I want to notice about this formulation of the dispute between realist and anti-realist is that it presupposes a prior inventory of recognitional capacities: a sector of reality can be judged recognition-transcendent, in any of the ways distinguished, only relative to antecedent assumptions about what faculties we have. And it is a notable fact that one realist strategy is to try to close the epistemic gap by claiming the existence of a faculty which crosses it. Dialectically the position is as follows: the realist formulates his conception of what the truth conditions of the given sentences consist in; the anti-realist protests that on that conception the truth conditions would objectionably transcend our faculties; the realist replies by disputing the assumptions about our faculties which underlie the anti-realist's protest, thus (as he hopes) restoring their accessibility. The debate is then apt to devolve upon the acceptability of the alleged faculty.

Examples showing this pattern are readily produced. Thus it has been claimed that the truth conditions of sentences about the past or about other minds are not, on a realist view, recognition-transcendent, because we can be credited with a perceptual capacity which (at least on some occasions) yields direct knowledge of the facts in question: so we do not have to picture our faculties vainly attempting, by means of some dubious process of inference, to traverse the gap at which the anti-realist jibs.[24] Somewhat so, a realist about ethical or mathematical sentences will find himself appealing to special cognitive faculties to account for our ethical or mathematical knowledge. Thus a mathematical platonist needs some account of how we come by mathematical knowledge, given his conception of mathematical reality: and here a faculty of mathematical 'intuition' has sometimes been introduced, its operations characterized in quasi-perceptual terms (Gödel 1964). And an ethical realist, locating values in

[24] This is John McDowell's (1978) view of the past and other minds.

166

the objective world, has to invoke a distinctive faculty of ethical apprehension directed upon such values.[25] In each of these cases we can say that the realist has introduced problematic cognitive faculties in order to link the facts as he construes them to our knowledge of those facts.

Actually, it is useful to distinguish two kinds of case. In one kind the form of the anti-realist complaint is that the realist truth conditions invite scepticism: this is because knowledge of those conditions appears mediated by a problematic inference – so with (e.g.) the past and other minds.[26] In the other kind the complaint is in a way more fundamental: the trouble here is that the anti-realist cannot comprehend how the introduced faculty is supposed to operate at all. That is, it is not that we have a clear idea of the mechanism of operation of the faculty but worry that it cannot reach far enough; rather, it is obscure what it would *be* for the alleged faculty to yield cognitive states consisting in a knowledge of the realist's truth conditions – so with (e.g.) abstract objects and ethical values. (As I shall later suggest, this difference turns upon the role of *causation* in the operation of the faculty.) So we have two ways in which the relation between knowledge and reality may be problematic on a realist conception: but they fall under a common rubric in that for both realism seems to require problematic epistemic faculties.

The formulation just arrived at differs in an important respect from Dummett's standard formulation. Dummett has the realist claiming recognition-transcendence in conjunction with bivalence. But, as we have seen, positions naturally classified as realist do not quite fit this description: for in some cases a realist precisely *denies* recognition-transcendence; he does not admit that his picture of truth conditions makes them unknowable. What he does is to secure this knowability by invoking problematic faculties. It therefore seems to me that, if we are to have a suitably general and uniform formulation of the realist/anti-realist dispute, we need to amend Dummett's official formulation in the way I have suggested.

The relevance of the foregoing remarks to the topic of modality is this: the shape realism takes with respect to modal sentences does not easily fit the mould Dummett casts; but it fits the broader formulation got by modifying Dummett's criterion in the suggested way. Its failure to conform to Dummett's characterization is shown in two (related) points. First, there is no class of sentences we can naturally identify as constituting *evidence* for modal claims and with respect to which the anti-realist proposes a reduction. That is, in Dummett's terminology (1968), the

[25] Thus ethical intuitionism. J. L. Mackie (1977: 28ff.) objects to precisely this feature of moral realism.
[26] Realism of this sort is discussed in my 'An *a priori* argument for Realism' (1979).

disputed class and the *reductive* class are not plausibly taken as standing in a relation of evidential support.[27] Second, realism about modality has not been accompanied by specifically *sceptical* worries: it has not been supposed that a realist is committed to viewing modal reality as stretching out beyond the evidence we can acquire about it. Rather, the focus of worry has been of the second kind I distinguished: the problem of what manner of faculty is capable of *any* sort of apprehension of modal facts. So we can expect modal realism to result in problematic faculties, not in scepticism. Anti-realism about modality will thus react to the problematic character of the needed faculty by denying that the truth conditions of modal sentences are as the realist claims: they will be said to consist instead in something (relatively) unproblematic. In other words, the anti-realist will propose a reduction of modal statements to others which do not present (or do not present the same) epistemological difficulties. What class of sentences will this reductive class be? The obvious answer is that it will consist of statements about what is *actually* the case. Thus anti-realism about modality is the doctrine known as *actualism*. For, as we shall see, actual facts are precisely those with respect to which modal facts are epistemically problematic. This is, of course, the traditional way in which the issue of modal realism has been set up. I arrived at this statement of the issue in a roundabout way because I wanted to show its bearing on the general formulation of realism, and to suggest how it naturally results from a prior general formulation derived from Dummett. Modal realism is correspondingly to be the thesis that modal facts are *not* reducible to actual facts.

A brief review of the traditional anti-realist positions concerning modality confirms their tacit actualism as well as their conformity to the epistemological criterion of modal anti-realism just suggested. Let us first distinguish two kinds of modal anti-realism, both of which have actualist presuppositions. By 'impersonal actualism' I mean views which try to reduce modal statements to statements about objective non-psychological conditions: thus causal necessity might be reduced to actual regularities in the world; logical necessity might be claimed to be reducible to statements about the actual world of a very high degree of generality; or dispositional statements to be reduced to categorical statements about the 'basis' of the disposition.[28] On the other hand, we have what might be called 'personal

---

[27] So it is not quite correct to characterize anti-realism as the thesis that truth always reduces to conditions of warranted assertion (evidence).

[28] That impersonal actualism is a form of modal anti-realism shows that truth in virtue of *external* facts is not sufficient for realism. Putnam (1975: 70) formulates realism in this inadequate way: anti-realism is not always idealist. Nor is such externality a necessary condition of realism, or else realism about one's own mental states would be impossible; indeed the anti-realist view here – namely behaviourism – consists precisely in reducing the subjective mental to what is objective and external.

actualism' (for want of a better label): this type of view suggests reducing modal statements to facts about the one who uses those statements; such a reduction is aptly described as psychologistic. Different versions of this general doctrine pick upon different sorts of properties of persons as constituting that to which a modal truth fundamentally reduces: thus the notions of stipulation or convention or intention or decision or imagination or mental disposition (Hume) are brought reductively to bear. On the former type of actualism, modality objectively resides in the impersonal world, but it does not transcend what actually obtains; on the latter, modality resides in us, either voluntarily or willy-nilly. In the case of the personal type of actualist reduction we are often presented with an interim reduction, i.e. a reduction of seemingly objective modal facts to psychological abilities or propensities. But since these psychological notions are themselves modally specified, the actualist owes us a further reduction of *those* if he is to complete his programme. The thesis of the modal realist is thus precisely that no such reductions are feasible: the modal is something 'over and above' what is merely actual.

This distinction between personal and impersonal anti-realist reductions has analogies in two other areas I have mentioned – namely, ethics and mathematics – which I will not take time to spell out, except to remark upon the very close analogy between Humean accounts of necessity and of value: both are conceived as objects of feeling, not of knowledge, arising from the *de facto* constitution of our minds.[29] These two anti-realist tendencies with respect to modality render modal knowledge (relatively) unproblematic, because they suggest familiar faculties by which modal truth, so reduced, may be known. The impersonal reduction brings modality within the scope of *perception* and extensions thereof; the personal reduction effectively assimilates modal knowledge to knowledge of one's own mental states, so that we know modal truths by something like *introspection*. By contrast, the realist view seems to place modality beyond the reach of such sublunary capacities, since modal facts cannot be identified with facts to which those faculties apply (I return to this below).

It is worth emphasis that the issue of realism about modality is not on the present conception an *ontological* issue; it is not, that is to say, an issue about what *objects* the world contains. Put in the formal mode, the issue relates, not to expressions whose semantic function it is to introduce entities into a sentence's truth conditions, but to expressions whose semantic role approximates rather to that of an *operator*. And the

---

[29] Hume is thus a non-cognitivist about both modality and value. See his *Treatise* Book I, Part III, Section XIV, on the idea of causal necessity; and Book III, Part I, Sections I and II, on the origin of ideas of value.

formulation of realism reflects this, since it speaks of the epistemological status of whole sentences and of what their truth consists in; indeed the notion of recognition-transcendence is best seen as applicable to sentences, not to singular terms. The question of realism concerns what kinds of *statement* we need in a complete and ultimate description of reality. (Compare the first sentence of Wittgenstein's *Tractatus*.) This point is significant for us in two ways. First, we see that arguments for the indispensability of a given kind of locution in our ultimate account of the world need not be taken as showing the inescapability of acknowledging entities of a certain kind: so the ontological commitments of a theory do not exhaust its commitments as to the ultimate facts it requires, i.e. the statements it has to take as true.[30] Second, construing the realism issue as ontological in character often unfairly prejudices the outcome against the realist view: for one is rightly reluctant to admit (e.g.) ethical or modal *entities* into one's ontology. If realism with respect to ethics or modality is taken as asserting that the world contains entitative moral values or possible objects, then it comes to seem totally unattractive; but there is no obligation to take it that way – it is fundamentally a thesis about the irreducibility of certain statements to others. And so no argument alleging the non-existence of such entities can be relevant to the question whether realism about such statements is correct. I suspect that a good part of resistance to modal realism in particular arises from tacitly assuming an ontological interpretation of the issue.[31]

I shall now sketch out three positions on the relation between the modal and the actual, endorse one of them, and finally indicate the epistemological consequences of the endorsed position. Since this position deserves to be called realist, the rest of the paper investigates the form and commitments of a plausible modal realism.

(i) Modal anti-realism is a thesis of reductive or eliminative actualism. That is, it claims that we do not need to recognize modalities in our fundamental account of what the world is *really* like: either we can offer adequate reductions of the import of modal statements, or we can simply refuse to indulge in such talk *ab initio*. But why should it be thought that modalities deserve our attention to begin with: what are the purposes for which we employ modal expressions, and are they legitimate? In other words, what is the (*prima facie*) utility of modal locutions?

There seem to be three main areas in which it is natural to invoke

---

[30] One is reminded here of Russell's (1956: 211ff.) defence of 'negative facts': his claim was that negative statements are needed in a full description of the world – and clearly this claim is not ontological in form, since it relates to the negation *operator*.

[31] This seems to lie behind the hostility to modal realism evinced in Mondadori and Morton 1976 (especially the second section). And I have heard similar objections to ethical realism.

modal notions: we invoke them in expression of our intuitions about the identity conditions of objects, as in philosophical discussions of essentialism; we invoke them in empirical scientific theories, in the notion of a law of nature and in ascriptions of dispositional properties; and we commonly appeal to necessity in characterizing the notion of logical implication. Since I myself take each of these areas seriously, I have a good deal of use for modal expressions. An actualist must therefore either refuse to set foot into these areas, or offer some actualist surrogate for the distinctions initially drawn in modal terms. The point is familiar from Quine: one cannot consistently help oneself to a kind of locution, ontological or otherwise, whose purport one officially repudiates: it must either be eschewed or reduced. An anti-actualist will thus make two claims: that modal locutions are indispensable in certain kinds of discourse themselves indispensable, and that they cannot successfully be replaced by any other sort of locution. The issue about modality is in close parallel with platonism in mathematics: we need mathematical expressions in scientific theories, and no programme of reinterpreting mathematical statements seems workable.[32] You may not like the resulting metaphysics, but it needs to be *demonstrated* that it is avoidable; so the onus is clearly on the actualist, or the anti-platonist.

To implement the actualist programme would therefore be to show that nothing of significance is lost if we purge our thought of all modal notions. This claim has always seemed to me totally implausible, since it seems to obliterate important distinctions between statements that we customarily and naturally draw in modal terms: between statements ascribing properties without which objects would not be what they are and those which ascribe properties that are merely accidental; between statements of law entailing counterfactuals and mere contingently universal generalizations lacking such entailments; and between inferences which are valid (necessarily truth-preserving) and those which are not. Of course, actualists have had things to say about these alleged distinctions, typically attempting to formulate them in non-modal terms. I cannot consider these various proposals here, but I think it is fair to report that the distinctions in question have proved remarkably difficult to reformulate in actualist terms, though this has often been disguised by covert circularities in the

---

[32] In fact, there may be a connection here as well as a parallel: for some philosophers, e.g. Putnam (1975) have suggested a modal interpretation of mathematics. There does, indeed, appear to be some trade-off between modality and abstract entities: logical consequence may be defined either model-theoretically, by quantifying over sets, or by use of modal operators; and the notion of possible world itself has been explicated in platonistic terms – as in Quine's (1969: 148ff.) construction of possible worlds in terms of mathematical structures of real numbers.

reductions offered.[33] If such reductions do indeed systematically fail, then I think we are committed to some sort of modal realism, given that we need to use modal locutions in contexts of serious assertion: we simply have no choice but to recognize a realm of irreducible modal facts. Indeed the failure of actualist reductions may be viewed as one more item on the list of defeats suffered by reductionist philosphers.[34] Nor is such failure altogether surprising: for if modal concepts really were just *equivalent* to concepts expressible in purely non-modal vocabulary, it would be a puzzle exactly why it was that we did not speak in that reductive actualist vocabulary to begin with. At any rate, we would need some explanation of why it is that modal expressions were introduced at all, if not to convey truths not expressible otherwise. (I offer this as a salutary thought, not as a conclusive argument.)

(ii) The denial of actualism, I have said, implies some form of modal realism, on the assumption that we regard modal sentences as true or false. Realism is the thesis that the truth conditions of the given class of statements transcend the truth conditions of statements of some relatively unproblematic (potentially) reductive class. But what is the exact character of this transcendence in the case of modality? A very natural answer to this question is that it consists in a relation of *independence* between modal truths and actual truths. Thus realism about the external world or about mental states plausibly consists in the thesis that the truth of such statements is not determined by the truth of experiential and behavioural statements, respectively.[35] It is a question whether such independence permits arbitrary conjunctions of material object and experiential statements, and of mental and behavioural statements, so that there are no mutual constraints whatever linking the concepts in question; but certainly a realist will insist upon a measure of independence in these cases which clearly precludes any possibility of reduction. This kind of independence thesis has the straightforward consequence that truths of the given class are not *supervenient* upon truths of the (potential) reductive class. So now, encouraged by this pair of examples, the anti-actualist might hold, similarly, that modal truths are independent of actual truths; in particular, the former do not supervene on the latter. Now supervenience is not an entirely perspicuous relation, but its minimal content will deliver with respect to modality the negative claim that two objects could be indiscernible with respect to non-modal predications yet differ with respect to what is modally true of them.

---

[33] Criticism of actualist theories of causal necessity can be found in Kneale 1949: §§13–20; Peacocke 1980; Stroud 1977: ch. x.

[34] Cf. a remark of Davidson's (1970: 91).

[35] See my 'An *a priori* argument for Realism' (1979) for more on this sort of realism.

Is this non-supervenience claim plausible? Hilary Putnam (1978: 164–5) makes some interesting remarks in this connection: he says, in effect, that reductive actualism is no more acceptable than phenomenalism about material objects, either for nomological modality or for strict logical modality. He expresses his 'modal-realist intuitions' by denying that 'what is true in possible worlds is totally determined by what is true in the actual world plus our conventions'. Thus he writes (1978: 164): 'does the totality of facts about what events actually take place determine the truth value of all statements of the form "it is possible that p"? To me, at least, it seems that the answer is "no", and if the answer is "no", then both Quinean accounts of logical necessity and Humean accounts of causality have to be wrong.' Putnam is, then, what might be called an independence realist about modality. We are not told how strong the failure of dependence is – whether what is modally true of an object is *completely* unconstrained by what is actually true of it – but we have enough to see that, for Putnam, modal truth may vary (*presumably* within some limits) while actual truth stays fixed.

This is certainly a statement of modal realism: but it does not seem to me that the independence thesis is plausible. Let us consider some cases. Suppose we have two samples of a given substance and suppose we have it that one sample has a certain dispositional property which generates certain counterfactuals about how it will behave in various possible circumstances; suppose also that the two samples have precisely the same microstructure as specified in non-modal physical vocabulary. Then I think we must say that the second sample has the same dispositional property and thus the same counterfactual properties as the first. But on Putnam's view the second substance might altogether lack that disposition, possessing instead some quite different disposition: but this seems to make nonsense of the idea that by investigating the actual microstructure of substances we can determine how they behave in counterfactual circumstances. (In fact, this is just the familiar point that dispositions have a 'categorical basis'.) Or again, consider two sectors of the universe in which the same sequences of (type) events occur: i.e. we have the same (type of) constant conjunction. Now if background conditions are (actually) the same and the events do not actually differ even down to their fundamental microstructural properties, then I cannot see how it might be that one sequence instantiates a law while the other does not: for surely if sequences of events differ in respect of lawlikeness that is due to some *actual* feature of the events concerned (or perhaps surrounding conditions).

So it seems to me that nomological modality does in this sense supervene on what is actually the case. Putnam may have been misled by his choice of example: he asks whether it is possible that there be two worlds indiscernible with respect to the occurrence of actual events in them but differing as to the fission of a small particle in a certain counterfactual experiment. So the question he is putting is whether two particles could be the same in all actual respects yet differ in their counterfactual properties. He makes no mention, in advocating an affirmative answer, of fundamental physical indeterminism, but perhaps this was at the back of his mind. Certainly, if quantum theory is to be believed, particles can behave differently though they be actually indiscernible; but we should note two points about this. The first is that it is extremely dubious that the truth value of the counterfactuals in question is determinate: for it does not seem that there is in this case a fact of the matter as to how the particle *would* behave were it subjected to certain energies; so we do not yet have a case in which the modal properties are both determinate and independent of actual properties. In fact, the indeterminacy of truth value of the counterfactuals here seems to trace precisely to the unavailability of any actual fact as a ground for the alleged modal difference. The second point to note is that such physical indeterminism cannot be supposed to account for *all* the cases in which Putnam would deny supervenience. For consider phenomena with respect to which such elementary indeterminism is irrelevant, or suppose that the world had been thoroughly deterministic: in these cases we would still want to make modal distinctions (perhaps based upon variations in initial conditions), but *ex hypothesi* these would not depend upon physical indeterminism.

What of the non-supervenience claim for strict or metaphysical modality? Consider first synthetic necessities, such as the necessity of origin, kind, composition and identity. Presumably the claim will take the form of envisaging two objects just alike in these respects – they instantiate the same non-modal properties and relations – yet for one object these properties and relations are essential while for the other they are not. So, for example, two human beings could both instantiate the relations that constitute having a certain origin – they developed in a certain way from gametes and so forth – yet one has his actual origin essentially while the other has his contingently; and similarly for the other synthetic necessities. To anyone who takes such essentialist claims seriously this must seem quite implausible: it is impossible that two objects be alike in their actual properties but differ in the modalities with which they instantiate those properties. Analytic necessities too seem supervenient upon actual facts. Two sentences could not agree in the actual meanings they have yet differ with respect to their necessity; nor could two sentences of the same logical form whose logical constants have the same meaning differ with respect to the modality of their truth value. So it seems to me that Putnam's independence formulation of modal realism is incorrect. I suspect he was induced to embrace it by rejection of actualism, which he rightly perceived to be anti-realist, and by the example of other areas in which independence *is* the proper expression of the realist view. However, I think there is a third intermediate view which avoids both actualist reduction and radical modal independence; to this view I now turn.

(iii) The position we have reached is this: we have two sets of concepts of which we wish to hold both that one set is irreducible to the other and that applications of concepts of the former set are not independent of applications of concepts of the latter set. Here we should be reminded of a parallel position with respect to the mental and the physical: it is arguable (indeed it seems to me correct) that we wish to draw genuine distinctions using mental vocabulary which cannot be reduced to distinctions drawn in purely physical terms, but that it is nevertheless not true that mental distinctions can apply independently of physical differences (Davidson 1970; McGinn 1980). The view on the relation between mental and physical concepts which captures this position is the thesis of supervenience without reduction: mental concepts differ in kind from physical, but there is no mental difference without a physical difference. Conscious sensations, as characterized by Thomas Nagel, are perhaps the most vivid illustration of properties which seem to meet these two conditions: since they are essentially subjective – and so differ categorially from objective physical states – there is no apparent prospect of successful reduction.[36] But still it seems that there is a sense in which the sensations a creature has are determined by its physical states. And this supervenience thesis seems to warrant the claim that, in some weak sense, any mental attribution is, in Dummett's phrase, 'true in virtue of' a physical fact, while not being reducible thereto.[37] A second case is that of ethical and descriptive concepts: it has been held that the ethical, while not reducible to the descriptive, nonetheless supervenes on it. Now it is not that such supervenience relations are completely unmysterious; but it does seem that a variety of considerations within the areas to which they have been applied strongly suggest that such supervenience is the only reasonable position.

I think that the modal and the actual constitute one more such case: there is some sense in which modal statements are true in virtue of non-modal or categorical statements, but this dependence is not so strong

[36] See especially Nagel, 'What is it like to be a bat?' (1979).
[37] Dummett (1976) also holds that counterfactual conditionals are true in virtue of categorical statements. It is unclear, however, whether he wishes this claim to carry reductionist commitments.

as to permit genuine reduction – on the contrary, modal statements mark out a real and irreducible range of facts. What is difficult, here as elsewhere, is to give an illuminating explication of the supervenience relation: to specify exactly *how* the statements in the domain of the relation determine the truth of statements in its range. Unfortunately, I have no very interesting suggestions to make along these lines: but, as Nagel says in another connection, one can know that something is true without yet knowing how it can be.[38] I am able to say, however, why it is that the supervenience claim is perfectly compatible with a realist view of modal truths as I have formulated realism. The reason is apparent from the following general principle: it is *not* true that if a class X of statements supervenes on a class Y of statements and a faculty F is proper to the acquisition of knowledge of statements of Y, then F is *eo ipso* proper to the acquisition of knowledge of statements of X. That is to say, the faculty F' proper to X may be epistemologically problematic relative to Y and F. Since different cognitive faculties may be needed in order to know the supervening truths from those adequate to the determining truths, it is possible to be a realist about the supervening truths: it is possible because our formulation of the transcendence definitive of realism was in terms of problematic cognitive faculties. Sensations and physical states well illustrate the above principle: according to Nagel, the capacity to know physical truths is possessible irrespective of one's own range of sensation types, but the capacity to know truths involving concepts for sensations is more narrowly constrained: one needs, to possess that capacity, to enjoy sensations similar to those the statement in question concerns.[39]

We could put this by saying that the name of the faculty needed to know a certain range of subjective truths is 'empathy': one may lack this faculty in respect of a class of subjective statements and yet be capable of coming to know the truth of statements which, as a matter of metaphysical fact, determine the truth value of the statements accessible only *via* the lacked faculty. If the determining truths constitute a potential anti-realist reductive class, then the need for an additional faculty shows that a realist interpretation of the supervening truths is being assumed: and this is precisely the situation with respect to the modal and the actual. (Some

[38] He says this of physicalism in 'What is it like to be a bat?' (1979: 176). If his suggestion in 'Panpsychism' (1979) is intended to explain how the physical can determine the mental, then *that* sort of suggestion will not go over to account for the supervenience of the modal and the actual ('panmodalism') for obvious reasons: and so it will just be a brute fact that objects have modal properties in virtue of their actual properties.

[39] Nagel does indeed remark that his view of mental states is strongly realist ('What is it like to be a bat?' (1979: 171)) since it implies the possibility of facts we can never, because of our very nature, comprehend. Clearly, this sort of realism involves facts which transcend our unproblematic recognitional faculties, and so fits my characterization of the realist position.

have supposed the same for ethics.) So the modal supervenes on the actual, but (as we shall see more fully) the modal is epistemologically problematic relative to the actual; we have then the defining characteristic of realism. Supervenience allows room for this transcendence because it is a rather weak relation between sets of statements: in particular, it does not imply that the supervening truths are *knowable via* the determining truths. Nor does it alone give an answer to the question of what the truth of the supervening statements *consists in*. It merely acknowledges a certain sort of non-contingent dependence. I think, therefore, that modal realism can be adequately formulated in terms of supervenience and does not require an independence formulation: for the definitive feature of realism is common to both sorts of formulation, namely transcendence from the recognitionally unproblematic. Moreover, this formulation of modal realism seems to me to give a more plausible account of the relation between modal and actual than the previous two. We have, then, what we sought: a non-objectual formulation of an acceptable modal realism.

I now wish to isolate and articulate what it is about modal realism (anti-actualism) that many philosophers, notably empiricists, find unpalatable. To do this is just to spell out the way in which modal statements are epistemologically problematic relative to actual statements. For I think that it is the epistemology of modality which is the source of the discomfort induced by modal realism. Dummett says at one place (1978: 169): 'We know what it is to set about finding out if something *is* true; but what account can we give of the process of discovering that it *must* be true?'[40] I am concerned in what follows to uncover the perplexity this question evokes. It is to be noted at once that anti-realist views of necessity have the effect of relieving the perplexity, as I remarked earlier: for modal truth conditions share the truth conditions of statements not supposed similarly problematic, since they do not transcend the realm of the actual. The thought I want to develop is that only what is actual is *empirical*, and that what is (to use a term of Richard Braithwaite's (1953: 318)) 'transempirical' is unknowable by means of the epistemic capacities we can intelligibly be supposed to possess. I shall concentrate on the case of causal or nomological modality.

[40] Here is perhaps an appropriate place to observe that the issue raised by this question applies equally to *de re* and *de dicto* modalities: for both we seem compelled to acknowledge a faculty which recognises modal properties, either of objects or of statements. It is therefore unclear how any reduction of all modalities to *de dicto* modalities removes the fundamental difficulty for empiricism – unless something further is said to actualise the *de dicto* modalities. Hume was particularly vulnerable on this point: for his critique of causal necessity can be recapitulated in respect of necessary relations among ideas, about which he has little positive to say. Strictly, he should adopt a non-cognitivist theory of the basis of our idea of necessity in the *de dicto* case too: subsequent empiricists have been similarly weak on this point.

A given linguistic construction may be called empirical if sentences in which it occurs can be (canonically) established as true by experience. Now the most obvious way in which the truth of a sentence can be empirically established is by *observation* – we can simply observe the sentence to be true. This method appears to require that we be able to observe that feature of the world which corresponds to the construction in question. Now what Hume (and his followers) famously insisted was that it is (necessarily?) false that we observe any feature of the world corresponding to expressions of causal necessity.[41] So if the content of a statement of law were stronger than the affirmation of an actual (though exceptionless and extensive) regularity, we could make no observation which would warrant the assertion of a statement with such a content. For we can observe only what is actual, and a statement of causal necessity purports to go beyond the actual; the causal necessity constructions are therefore transempirical in the sense that they fail to yield observably true sentences.

All this seems to me correct, but as a reason against objective modalities it is totally unsatisfactory; for the criterion of empirical significance it assumes – direct observability – is much too strong. More exactly, it indiscriminately brands various constructions as non-empirical which are not plausibly so taken, and it thus obliterates an intuitive epistemological difference we should like to capture between modality and these other cases.[42] Three sorts of expression may be cited: spatial, temporal and theoretical. It does not seem true that places and times are empirically inaccessible in the sense specified, yet they are hardly *observable* entities. (This can be seen by asking how a causal theory of perception might be applied to such entities.) If modality were no worse off epistemologically than space and time (even on an absolutist theory of them), then I think a reasonable empiricist would have no cause for complaint about modal realism: we need to find something more distinctively transempirical

---

[41] Thus A. J. Ayer (1972:4) writes, 'There is no such thing as seeing that *A must* be attended by *B*, and this not just because we lack the requisite power of vision but because there is nothing of this sort to be seen.' And J. L. Mackie (1974: 217) asks, with respect to Kneale's account of nomic necessity, 'What, in the operation of such a mechanism, however delicate and ingenious, could we see except the succession of phases?' The defect of both formulations, inherited from Hume, is the use of the verb 'to *see*': below I try to formulate the Humean worry in less naive terms.

[42] Two writers who underestimate the distinctively non-empirical character of modality are F. P. Ramsey and Richard Boyd. Ramsey (1931), in his last papers, says that 'realism' (his term) about causal modalities and about theoretical entities is equally unplausible in view of the observation-transcendence they both imply: pp. 253, 261. Boyd (1976), on the other hand, being already a realist about theoretical entities, remarks that, since the considerations in the two areas are so similar, we should take on modal realism as well. But, as I shall suggest, it seems to me crucially important that there is a sense in which modality and theoretical unobservables differ in point of their empirical status.

about modality. The postulation of causally operative unobservables in physical science also generates sentences which are not strictly observably true: but again, it is not the case that the truth value of such sentences is empirically unascertainable. This is because the theoretical postulation of unobservables has observable empirical *consequences* which permit the theory to be tested by experience.

So the question a Humean about modality has to answer is how modal realism differs from realism about theoretical entities or about space and time on the score of empirical significance. That is, what version of empiricism is liberal enough to count these other cases empirically significant but restrictive enough to give a sense in which modality is objectionably transempirical? To answer this question would be to show what is epistemologically problematic about modal realism. Unless this can be done Humeans are vulnerable to the charge of resting upon a naive and oversimple conception of the conditions of empirical knowledge.[43] The charge will be put in this way: knowledge arrived at by inference to the best explanation of empirical phenomena is itself genuinely empirical knowledge; and it appears that modal notions *are* bound up in our empirical theories of the world – counterfactuals, dispositions, lawlike necessity, and so on. In short, we can empirically know nomically modal truths by inference to the best explanation of observable empirical facts. Now I think that there is something crucially wrong with this suggestion, but it is not altogether obvious how the intuitive epistemological difference between modal constructions and the others mentioned is to be formulated. I shall try to bring out the difference by borrowing some ideas suggested by Hartry Field's (1980) treatment of mathematics.

Field claims (and claims to prove) that mathematical theories are *conservative* in the following sense: if you take a nominalistic theory N and add to it a mathematical theory M (as when applying some pure mathematical theory to a range of physical phenomena), then the resulting theory $N \cup M$ is a conservative extension of the theory N – i.e. there are no nominalistic consequences of $N \cup M$ which were not already consequences of N. Field further suggests how the axioms of a scientific theory might be nominalized, so that the empirical assertions of the theory do not themselves contain mathematical vocabulary. Granted both of these claims about the role of mathematics in scientific theories, Field uses them to undermine Quinean indispensability arguments for platonism, and so paves the way for a fictionalist account of mathematical sentences. But I want to appropriate these points about mathematics to show a different thing. Suppose one is a platonistic realist about mathematics,

[43] Putnam (1975: 69) registers this kind of complaint, and Bas van Fraassen urged the point in discussion.

and suppose that one agrees that the application of mathematics to a nominalized theory has the stated conservativeness property. Then the empirical consequences of a physical theory, these being nominalistic, will follow even without the use of mathematics in the theory, since the theoretical axioms are nominalistically formulable and mathematics is conservative. But if so, it is implausible to hold that the theory $N \cup M$ is known by way of the empirical consequences of that theory, since these are equally consequences of N. That is, under a platonist view, $N \cup M$ is a factually stronger theory than N – it asserts more truths – yet the two theories are empirically equivalent: how then could we claim that the mathematical component of the theory is known by inference to the best explanation of the empirical consequences of the theory? For the empirical evidence is, by conservativeness, compatible with a much weaker theory: mathematics does not increase empirical content.

I think this shows that a certain strategy for combining mathematical realism with empiricist epistemology – to the effect that since mathematical sentences occur in empirical theories they are known, like the nominalistic portions of such theories, by inference to the best explanation – is inadequate, because of the essentially non-empirical role that mathematics plays. In this respect of conservativeness mathematical entities and theoretical entities stand in striking contrast: for sentences about theoretical entities are *not* nominalistically conservative – they issue in genuinely new empirically testable consequences. The lesson for us in this contrast is this: though there is a clear sense in which numbers and theoretical entities are unobservable, they yet differ radically in point of empirical significance. And this difference suggests that there is no prospect of putting platonistic mathematics on an empirical footing; our knowledge of mathematical truths (given that there are any) *is* therefore a threat to thoroughgoing empiricist epistemology, despite their occurrence in (incontestably) empirical theories.[44] We can now state a usable criterion of the transempirical: a given type of construction gives rise to transempirical truths if (a) it is interpreted realistically and (b) its introduction into an empirical theory yields a conservative extension of that theory (or equivalently, its removal from an empirical theory leaves its empirical consequences intact). Mathematics is transempirical by this criterion, whereas statements about theoretical entities are not. The question now is whether modality is transempirical in this sense.

---

[44] I take this consideration to undermine the account of mathematical knowledge proposed by the authors cited in n. 10, thus reinstating the epistemologically problematic status of mathematical truths, as platonistically interpreted.

Let T be a theory free of modal expressions: its generalizations apply only to all actual objects of the kind treated by the theory, past, present and future; and suppose we know T to be true, presumably by observation and induction. (We can think of T as got by taking a scientific theory containing modal expressions, implicitly or explicitly, and removing its modal content.) T will have certain empirical consequences which are used to verify it. Now add to T some causally modal constructions V – a nomic necessity operator, a dispositional suffix, a counterfactual conditional (possibly embedded in a modal logic appropriate to the modal notions introduced) – and suppose that the resulting theory $T \cup V$ is realistically true. Then the modalized theory $T \cup V$ is, for a modal realist, a factually stronger theory than the non-modal T: it reports, not just what actually happens, but what happens in all causally possible conditions. But clearly $T \cup V$ is a conservative extension of T: in particular, $T \cup V$ has no empirical consequences not shared by T. So modality, like mathematics, is empirically conservative. The reason is obvious: empirical consequences are reported by sentences which can be observed to be true, but what is non-actual cannot be observed to be true. To put it (heuristically) in terms of possible worlds: the modalized theory says what goes on in nomologically possible worlds, but empirical consequences obtain only in the actual world. So, by the conservativeness of modality, removing the modal component from an empirical theory does not decrease empirical content. In fact, this is just the point which has seemed to empiricists to lend such support to a constant conjunction conception of laws: viz., that the extra assertoric content alleged by the necessitarian about laws must transcend what we can empirically verify.

It follows, I think, that we could not plausibly be said to come to know a theory with such modal content by purely empirical means: for the empirical consequences of the theory are compatible with a weaker theory in which modalities do not figure.[45] So in this crucial respect, captured by the idea of conservativeness, modality and mathematics are alike in not being properly empirical, and different from other sentences whose subject matter is admittedly not itself observable, e.g. theoretical

---

[45] Here we must guard against a mistaken reaction to this claim. It may be said: 'knowledge of modalities is bound up with our inferential knowledge that certain generalizations are *explanatory*, since explanation is only by laws whose necessity sustains counterfactuals: so knowledge of modality does result from inference to the best *explanation* of the empirical consequences of a theory'. However, this objection tacitly concedes the point I am after: for now we will need to appeal to knowledge of the lawlike status of generalizations in the *premisses* figuring in the inference – whereas the attraction of the original proposal was to give modal knowledge a purely empirical (actualist) basis. (This point will become clearer when I give a more positive account of our knowledge of modality.)

sentences.[46] And this suggests a restriction upon the propriety of epistemological theories of a given type of statement which are based on the idea
of inference to the best explanation: namely, that the statements in
question should not be empirically conservative. If this is correct, the
liberalized empiricist can insist upon the non-empirical nature of modality
while not simultaneously and unwantedly excluding statements whose
epistemological credentials he finds (or should find) acceptable. But now
if modality is thus non-empirical and if modal realism *is* (as I have
claimed) true, then this insistence immediately refutes empiricism, be it
ever so attenuated: for that combination of views is tantamount to the
thesis that there are facts, about which we can apparently know, that are
not epistemically accessible to us through faculties acceptable to a
consistent empiricist.[47] And of course this is essentially just the predica-
ment that Hume perceived, only he chose to reject objective necessities.

How then *does* one know causally necessary truths? Or better: how does
one know *of* a scientific generalization that it is a nomological necessity?
Since it does not seem right to invoke recognition of a relation of
confirmation between empirical data and generalizations alone, it appears
that we need to introduce a faculty directed upon *intrinsic* features of the
content of a lawlike generalization. I tentatively suggest that knowledge
that a given generalization is a law is a result of the operation of
recognitional faculties of two fundamentally different sorts: there is first
the faculty that tells us, from the intrinsic properties of the generalization,
whether or not it is law*like*; and there is second the faculty that allows us to
confirm the generalization by empirically recognizing instances of it.
Indeed, the verification of causal necessities seems to follow the pattern set
by the structure of our knowledge of *a posteriori* metaphysical necessities,

---

[46] I do not want to suggest that mathematics and modality figure in empirical theories in
exactly analogous ways; indeed, I think there is an important point of disanalogy to be
noticed. Field (1980) argues, plausibly, that the role of mathematical statements in
scientific explanations is *extrinsic* to the phenomena being explained: and this is an
important part of his case for adopting a fictionalist attitude towards mathematical
sentences. However, it seems that modality is not thus extrinsic to explanations: for one
does not have a genuine explanation unless modalities are implicated – so it would not be
feasible to combine a realist account of scientifically explanatory sentences with fictionalism about their modal status.

[47] The need for non-empirical cognitive faculties was, of course, insisted upon by the
Rationalists. Thus Leibniz, e.g. in the 'Introduction', to *New Essays on the Human
Understanding*, remarks that our knowledge of 'necessary truths, such as we find in pure
mathematics, and particularly in arithmetic and geometry, must have principles whose
proof does not depend upon instances, nor consequently on the testimony of the senses',
since the necessity of a generalization cannot be established merely by observing its actual
instances. And Descartes, in the second *Meditation*, tells us that the dispositional properties
(which are implicitly modal) of (a piece of) wax can be known only by reason, not by
perception or imagination.

outlined earlier. Knowledge that a generalization is lawlike is knowledge that if true it is a law (i.e. necessarily true), and in the case of metaphysical necessities we likewise have a conditional premiss with a modal consequent. This is then conjoined with a purely empirical non-modal premiss, affirming the truth of the antecedent. The modal claim – nomological or metaphysical – then follows by detachment. In the case of metaphysical necessities we know the conditional premiss by appreciating that possession of the property in question is constitutive of an object's identity; in the case of laws we judge a generalization lawlike if (roughly) the predicates in it stand in some explanatory and projectible relation to each other: and both pieces of knowledge are possessed in advance of knowing whether the objects in question in fact (actually) instantiate the properties ascribed in the non-modal premiss.[48] So nomological and metaphysical modality seem, upon examination, to have a (somewhat) uniform epistemology. This reflection prompts the following general thesis: there is a clear and important sense in which *all* specifically modal knowledge is *a priori*. It is not surprising, if that thesis is true, that modality, like mathematics, is empirically conservative: anything true *a priori* is bound to be.

The above thesis puts modal realism in a new light. For it suggests that the problematic character of modal knowledge, as implied by my formulation of realism in terms of cognitive faculties, traces to its *a priori* status. A faculty yielding *a priori* knowledge has always seemed to philosophers difficult to comprehend. Knowledge of the actual empirical world is arrived at *a posteriori* and is *pro tanto* unproblematic; but if the transcendence of the modal over the actual calls for an *a priori* faculty, we see at once in what the problematic character of modal knowledge consists. It is just the old problem of how *a priori* knowledge is possible. If so, the problem of modal knowledge reduces to a more general problem. I do not, alas, have a satisfactory solution to the general problem, but I think I can at least formulate it sharply enough to see the nature of the underlying issue.

It is plausible and illuminating to define the distinction between *a priori* and *a posteriori* truth along the following lines: an *a posteriori* truth is one that must be known by way of causal interaction with the subject matter of some justifying statement; and an *a priori* truth is one that can be known without causal interaction with the subject matter of some justifying

---

[48] I am aware that this account of our knowledge of nomic necessity is sketchy and obscure – also that there are important differences between the epistemic status of the two sorts of modal conditional. It does, however, seem to me desirable to find something uniform in the epistemology of all types of modality.

statement.[49] Now if modal statements are indeed known *a priori* we can say (simplifying a bit) that the cognitive states which constitute knowledge of modal statements are not *caused* by the modal facts in virtue of which the known statement is true. This immediately gives the consequence that modality is not perceptible: we cannot perceive modality (have an 'impression' of necessity) because modal knowledge is *a priori*, and *a priori* knowledge is not by definition based on the kind of causal process involved in exercises of the faculty of perception. But note that the definition of the *a priori* does not tell us how *a priori* knowledge *is* acquired: the characterization is entirely negative. Of course names of appropriate *a priori* faculties are not far to seek: 'reason', 'intuition', and the like. But these do not afford any real hint as to the *mechanism* or mode of operation of the faculties denoted. The point can be put generally and intuitively as follows: our conception of knowledge – that is, of the relation between knowledge and reality – construes the state of knowing as somehow the *effect* of that which is known. Thus perceptual knowledge is our basic model of how knowledge comes about (the causal theory of knowledge is built upon this model): and we conceive of other kinds of knowledge – in memory or by induction – as approximating more or less closely to this model. But with *a priori* knowledge the model seems to break down altogether. Either we try to conceive of a non-causal mode of influence upon the knowing mind, which seems incoherent; or we decide to give up the idea that knowledge somehow results from what is known, which leaves us perplexed about what such knowledge consists in and in want of an alternative conception. Empiricists are notoriously hard put to account for knowledge that seems inescapably *a priori*, and the naturalistic conception of knowledge in terms of causality seems equally impotent to account for it.[50]

---

[49] See my 'A *priori* and *a posteriori* knowledge' (1976). The proposed definition improves on the traditional characterization in terms of 'independence from experience' in a number of respects. (i) It is unclear what notion of experience is being employed in the traditional characterization: it cannot allude to the phenomenological kinds enjoyed by human beings, since we would wish to apply the *a priori/a posteriori* distinction to creatures whose experiences were phenomenologically dissimilar – and a more general definition will, I think, resort to causal notions; (ii) the traditional characterization wrongly pronounces introspective knowledge *a priori*, since we do not have experiences of our experiences, while the causal definition seems capable of handling this type of *a posteriori* knowledge; (iii) the distinction, or an analogue of it, seems applicable to informational states in whose production experience can play no part, e.g. a recording device that acquires information about the environment by means of 'sensors', and also generates informational states, such as the state of having 'proved' a mathematical theorem, on the basis of a mathematical programme. But I cannot elaborate on these points now. (It is interesting to note that Russell (1973: 42ff.) held ethical knowledge to be *a priori*, analogously to our knowledge of necessity.)

[50] We now have enough before us to make some brief remarks on Peacocke's (1980) discussion of the empirical status of causal modalities. Peacocke suggests that his

The epistemological problem with modality is, then, that we cannot represent modal facts as causally explaining our knowledge of them. And the trouble with this is that we seem to have no other going theory of knowledge. We thus reach the uncomfortable position of agreeing that there is *a priori* knowledge but not understanding how such knowledge comes about. And this, it seems to me, is the form that the problematic epistemology of modal realism takes. (No such difficulty afflicts the realist about space and time.) My own view is that we are here confronted by a genuine and intractable conflict between what our metaphysics demands and what our epistemology can allow.[51] If modal realism is to be finally

---

anti-actualism does not make causal modalities non-empirical because (a) we employ modal notions in constructing empirical theories, and (b) we can give an empirically checkable manifestation condition for possession of beliefs involving causal modalities (p. 64). We should note two points about these claims. The first is that suggestion (a), as it stands, seems to render mathematics similarly empirical: but we have seen for this case, as well as for modality, that there is a very significant difference between the epistemic status of these statements and that of, e.g. statements about theoretical entities. The second is that there seems no obvious reason why the availability of an empirical manifestation condition should show the notion manifested to be *itself* empirical, i.e. such that truths involving it can be known by purely empirical means: again, think of mathematics. Peacocke also compares the causal necessity operator with the universal quantifier, and finds no good reason to regard them, as philosophers traditionally have, as importantly different in respect of empirical status (p. 59). Here we should distinguish three claims: that universal quantification over infinite domains is problematic in much the way that causal modalities have been supposed to be; that universal quantification over (humanly) unsurveyable domains is similarly problematic; and that universal quantification itself, irrespective of the surveyability of the domain, is epistemically comparable to modality. About these three claims I would say the following. The first claim is true, but because of special problems about infinity, not because of the sense of the universal quantifier itself. The second claim ignores the fact that quantitative extensions of ordinary empirical faculties would be enough to make such unsurveyable domains empirically accessible, but the move from the actual to the possible requires a qualitatively different epistemology, since modal facts are empirically conservative. The third claim is reminiscent of the thesis, maintained by Ramsey and Russell, that general facts are irreducible to sets of particular facts because you always need to add some quantified condition to any set of singular statements if you are to capture the import of an entirely general truth. The issues here are fairly subtle, but I think we can point to two important disanalogies between this irreducibility thesis and modal realism. First, the assertibility conditions of universally quantified statements seem statable by means of purely singular statements; whereas, as we have seen, causally modal truths cannot be known on the basis of purely actualist premises. Second, and related, one can check, for directly inspectable domains, by perfectly empirical means that *all* of the objects in question have been covered in the set of verifying singular statements, but one cannot thus *empirically* check whether a universal generalization holds in all *possible* circumstances. So I think that Peacocke underestimates the epistemological problems raised by modal realism. (I am, however, in considerable agreement with much else in his paper.)

[51] This tension between the metaphysics of modality and the requirements of an intelligible epistemology is, of course, precisely analogous in form to that described by Benacerraf (1973) in respect of mathematics.

accepted, it must find some way of alleviating the conflict to which it gives rise. My aim has been to articulate in what the difficulty ultimately consists.[52]

REFERENCES

Adams, R. M. 1974. 'Theories of actuality.' *Nous* 8, 211–31.

Ayer, A. J. 1972. *Probability and Evidence*. London: Macmillan.

Baldwin, T. 1975. 'Quantification, modality and indirect speech.' *Meaning, Reference and Necessity*, ed. S. Blackburn, pp. 56–108. Cambridge: Cambridge University Press.

Benacerraf, P. 1973. 'Mathematical truth.' *Journal of Philosophy* 70, 661–79.

Boyd, R. 1976. 'Approximate truth and natural necessity.' *Journal of Philosophy* 73, 633–5.

Braithwaite, R. 1953. *Scientific Explanation*. Cambridge: Cambridge University Press.

Davidson, D. 1967a. 'The logical form of action sentences.' *The Logic of Decision and Action*, ed. N. Rescher, pp. 81–95. Pittsburgh: University of Pittsburgh Press.

Davidson, D. 1967b. 'Truth and meaning.' *Synthèse* 17, 304–23.

Davidson, D. 1969a. 'The individuation of events.' *Essays in Honor of Carl G. Hempel*, ed. N. Rescher. Dordrecht: Reidel.

Davidson, D. 1969b. 'On saying that.' *Words and Objections*, ed D. Davidson and J. Hintikka, pp. 158–74. Dordrecht: Reidel.

Davidson, D. 1970. 'Mental events.' *Experience and Theory*, ed. L. Foster and J. Swanson, pp. 79–101. London: Duckworth.

Davies, M. 1978. 'Weak necessity and truth theories.' *Journal of Philosophical Logic* 7, 415–39.

Dummett, M. 1968. 'The reality of the past.' *Proceedings of the Aristotelian Society* 69, 239–58.

Dummett, M. 1973. *Frege: Philosophy of Language*. London: Duckworth.

Dummett, M. 1976. 'What is a theory of meaning? II.' *Truth and Meaning*, ed. G. Evans and J. McDowell, pp. 67–139. Oxford: Clarendon Press.

Dummett, M. 1978. *Truth and Other Enigmas*. London: Duckworth.

Field, H. 1980. *Science without Numbers: A Defence of Nominalism*. Oxford: Blackwell.

Gödel, K. 1964. 'What is Cantor's continuum problem?' *Philosophy of Mathematics: Selected Readings*, ed. P. Benacerraf and H. Putnam, pp. 258–73. Englewood Cliffs, N. J.: Prentice Hall.

Gupta, A. 1978. 'Modal logic and truth.' *Journal of Philosophical Logic* 7, 441–72.

Hart, W. D. 1979. 'Access and inference.' *Proceedings of the Aristotelian Society*, supp. vol., 153–65.

Kneale, W. 1949. *Probability and Induction*. Oxford: Oxford University Press.

Kripke, S. 1963. 'Semantical considerations on modal logic.' *Acta Philosophica Fennica* 16, 83–94.

[52] I am grateful for comments on the ideas developed in this paper to members of the UCLA philosophy department and to participants in the Thyssen group meeting in 1979: Anita Avramides, Hartry Field, Bas van Fraassen, Bill Hart, Christopher Peacocke and Warren Quinn were particularly helpful.

Kripke, S. 1965. 'Semantical analysis of intuitionist logic 1.' *Formal Systems and Recursive Functions*, ed. J. N. Crossley and M. Dummett, pp. 92–130. Amsterdam: North Holland.

Kripke, S. 1972. 'Naming and necessity.' *Semantics of Natural Language*, ed. D. Davidson and G. Harman, pp. 253–355. Dordrecht: Reidel.

Lewis, D. 1968. 'Counterpart theory and quantified modal logic.' *Journal of Philosophy* 65, 113–26.

Lewis, D. 1970. 'Anselm and actuality.' *Nous* 4, 175–88.

Lewis, D. 1973. *Counterfactuals*. Oxford: Blackwell.

McDowell, J. 1978. 'On "The Reality of the Past".' *Action and Interpretation*, ed. C. Hookway and P. Pettit, pp. 127–44. Cambridge: Cambridge University Press.

McGinn, C. 1976. '*A priori* and *a posteriori* knowledge.' *Proceedings of the Aristotelian Society* 77, 195–208.

McGinn, C. 1979. 'An *a priori* argument for realism.' *Journal of Philosophy* 76, 113–33.

McGinn, C. 1980. 'Philosophical materialism.' *Synthèse* 44, 173–206.

Mackie, J. L. 1974. *The Cement of the Universe*. Oxford: Clarendon Press.

Mackie, J. L. 1977. *Ethics: Inventing Right and Wrong*. Harmondsworth, Middx.: Penguin.

Mondadori, F. and Morton A. 1976. 'Modal realism: the poisoned pawn.' *Philosophical Review* 85, 3–20.

Nagel, T. 1979. 'What is it like to be a bat?' 'Subjective and objective.' 'Panpsychism.' *Mortal Questions*. Cambridge: Cambridge University Press.

Peacocke, C. 1976. 'An appendix to David Wiggins' "Note".' *Truth and Meaning*, ed. G. Evans and J. McDowell, pp. 313–24. Oxford: Clarendon Press.

Peacocke, C. 1978. 'Necessity and truth theories.' *Journal of Philosophical Logic* 7, 473–500.

Peacocke, C. 1980. 'Causal modalities and realism.' *Reference, Truth and Reality*, ed. M. Platts, pp. 41–68. London: Routledge and Kegan Paul.

Prior, A. N. and Fine, K. 1977. *Worlds, Times and Selves*. London: Duckworth.

Putnam, H. 1975. 'What is mathematical truth?' *Mathematics, Matter and Method (Philosophical Papers*, vol. 1), pp. 60–78. Cambridge: Cambridge University Press.

Putnam, H. 1978. 'There is at least one *a priori* truth.' *Erkenntnis* 13, 153–70.

Quine, W. V. 1961. *From a Logical Point of View*. New York: Harper and Row.

Quine, W. V. 1969. 'Propositional objects.' *Ontological Relativity and Other Essays*, pp. 139–60. New York: Columbia University Press.

Quine, W. V. 1979. 'Intensions revisited.' *Perspectives in the Philosophy of Language*, ed. Peter A. French *et al.*, pp. 268–74. Minneapolis: Minnesota University Press.

Ramsey, F. P. 1931. *Foundations of Mathematics and Other Essays*. London: Routledge and Kegan Paul.

Rescher, N. 1975. *A Theory of Possibility*. Oxford: Blackwell.

Russell, B. 1956. 'The philosophy of logical atomism.' *Logic and Knowledge*, ed. R. C. Marsh, pp. 175–282. London: Allen and Unwin.

Russell, B. 1973. *Problems of Philosophy*. Oxford: Oxford University Press.

Stalnaker, R. 1976. 'Possible worlds.' *Nous* 10, 65–75.

Steiner, M. 1975. *Mathematical Knowledge*. Ithaca, N.Y.: Cornell University Press.

Stroud, B. 1977. *Hume*. London: Routledge and Kegan Paul.

Essences and laws of nature[1]

BAS C. VAN FRAASSEN

Robert Silverberg's science fiction novel, *The Stochastic Man*, begins with reflections on some aspects of a view I wish to defend. The first few lines are:

> We are born by accident into a purely random universe. Our lives are determined by entirely fortuitous combinations of genes. Whatever happens, happens by chance. The concepts of cause and effect are fallacies. There are only *seeming* causes leading to *apparent* effects. Since nothing truly follows from anything else, we swim each day through seas of chaos, and nothing is predictable, not even the events of the very next instant.
> Do you believe that?
> If you do, I pity you, because yours must be a bleak and terrifying and comfortless life.

Despite all the distinctions we philosophers would cautiously draw, we easily recognize the rejected view as the nominalist–empiricist tradition of Nicholas of Autrecourt, William Ockham and David Hume – and indeed, as the view of a rather better known literary protagonist, Sartre's Antoine Roquentin.

The aspect of this view which Silverberg's narrator specifically denigrates is the denial of necessary connections among events, necessities in the natural order, laws of nature, determinism in all or part of the world. While it does not really follow from this view that there cannot be successful predictions, it does follow that no prediction has its success guaranteed by the way things *must* turn out, nor are any actual regularities explicable on the basis of necessary concomitance. The denial extends to causality, dispositions and objective chance or propensity. The associated view of science is that its central topic of concern is exactly what is concrete, actual, and observable, and that all else is to be located in the conceptual resources mobilized to model the subject of that central concern.

I shall examine some questions relating to this debate, and argue

[1] This paper was written during a leave of absence funded by a University of Toronto Connaught Senior Research Fellowship, which is gratefully acknowledged.

concerning essences, laws of nature, and counterfactual conditionals that the supposed objective modal distinctions drawn are but projected reifications of radically context-dependent features of our language.

I THE WARRANT FOR A COUNTERFACTUAL CONDITIONAL

The modern problem of counterfactuals was introduced by Goodman and Chisholm in the 1940s. It was generally observed at that time that the warrant for such a conditional assertion is generally, or at bottom, provided by a law of nature. This observation was the starting point for such partial theories of these conditionals as were framed during that decade and the next by such philosophers as these, Reichenbach and Sellars. Examples spring readily to mind:

> (1.1) If this nickel (coin) were heated to n degrees Celsius it would melt, *because* the melting point of nickel is n degrees Celsius.

Merely universal generalizations cannot play this warranting role. The paradigm example was the contents of Goodman's pocket on V.E. day. All the coins in it were copper, whose melting point is q; but 'If this nickel coin had been in my pocket on V.E. day, it would have melted if heated to q degrees Celsius' is not warranted. On the other hand, the mere falsity of the antecedent does not refute it. It was concluded that counterfactual conditionals which are true, are so because of laws of nature, and not because of merely accidental universal truths.

It will spring to the eye that the law cannot be sufficient warrant for the counterfactual all by itself. Generally accepted, and apparently operative here, is the inference (provided being F does not contravene physical possibility):

> (1.2) It is a physical law that all F are G; hence, if *this* were F it would be G:
> (1.3) $\boxed{P}$ (A $\supset$ B); therefore A $\rightarrow$ B

where $\boxed{P}$ signifies physical necessity, and $\rightarrow$ is the proper conditional connective. Now it is presumably physically necessary that if something is nickel, then it will melt if heated etc. But if our only additional premise were the fact that this coin is nickel, we would be guilty of the modal fallacy:

> (1.4) Fa; $\boxed{P}$ (Fa $\supset$ (Ga $\supset$ Ha)) therefore $\boxed{P}$ (Ga $\supset$ Ha) and hence also Ga $\rightarrow$ Ha.

At this point one sees the temptation to give up the program for warranting conditionals by appeal to physical laws, and to say that the relevant part of our knowledge of natural necessity is the general conditional (x) (Fx → (Gx → Hx)). But that is the counsel of despair; much easier, and quite traditional, is to locate the missing premise among attributions of essence:

(1.5) This coin is essentially nickel, that is, it is nickel in every possible world (and hence, in every physically possible world) in which it exists.

which supplies the premise ℙ Fa provided this is understood with the relevant existence qualification which speakers may be assumed to leave tacit: 'This coin would melt in any circumstances in which it [exists and] is heated to n degrees, *because* this coin is necessarily made of nickel [if it exists at all], and all things made of nickel necessarily melt if heated to n degrees.'

2 ESSENCE

It is very interesting to note this intimate connection between singular counterfactuals and essences, for it provides a practical way to ferret out our opinions about essence. Since Locke it has been a familiar position that the mechanical, chemical, molecular or atomic structure of a thing constitutes all or part of its essence. We may now see corroborating evidence for this opinion in our willingness to insist that any coin would have the same melting point regardless of whether it had been in Goodman's pocket, or indeed, in many other situations. Much will depend of course on how we complete the general theory of essences.

One basic question for such a general theory must be whether the laws of nature of a world automatically contribute to the essence of things in it. Suppose that this coin is essentially nickel, that is, it is nickel in all possible worlds in which it exists. In addition, suppose that it is physically necessary that nickel melts at n degrees. It follows that if this coin were (to exist and) have that temperature, it would melt. But does it follow that this coin has, not just physically necessarily, but essentially, the property that it melts if heated to that point?

First, we can frame sensible counterfactuals whose antecedents deny actual laws of nature. David Lewis coined the term 'counterlegals' for them. Thus we can say 'If the world had been Newtonian, . . .' or 'If the melting point of nickel had been q degrees, . . .' Second, at least at first sight, we do not seem to assume automatically that a thing in this world would not exist if some law were lifted or violated. Thus I say, 'If the strength of the gravitational field were to diminish drastically, due to a previously unknown time dependence of the gravitational constant, we should certainly notice: we should all float to the ceiling.' That conditional may be true or false, but it is not senseless, nor is there any *prima facie* reason to think that the word 'we' occurring in it does not denote us.

To return to our question, then, let us consider both alternatives. Suppose to begin that the physical necessities of this world are not automatically part of the essences of the things in it. In that case, prospects look bleak for the empirical inquiry into what properties do belong to the essence of anything. For I can set up experiments that realize physically possible conditions, such as heating nickel objects to a certain temperature. These experiments will support general hypotheses about the laws of nature – at least, that is the idea of those who believe in them. Assuming also that this object is necessarily or essentially nickel, I will reach conclusions about the properties it has with physical necessity. But if essence is independent of physical necessity, I won't know at all yet whether those properties are also essential. Hence the route of empirical inquiry is blocked. We might define the *sub-essence of* x *in a world* as the totality of properties implied, via the physical laws holding in that world, by its essence. Empirical access is then possible directly to the sub-essence only.

The other alternative, that if some property is physically necessary in this world, then it belongs to the essence of everything in it, is therefore much more appealing. It removes that supposed limit to experimental knowledge about the essence.

And indeed, I have found, upon reflection, the most marvellous possibilities for a physics based on essences if that alternative is embraced. As is well known, our world is not deterministic. This raises special problems for causal explanation for us.[2] Let us imagine that in a certain world, physically possible relative to ours, the irreducibly statistical law for the melting point of nickel is that it is always in the interval n ± d degrees Celsius, with a uniform probability distribution on that interval. It is quite conceivable that certain objects provide an exception to this law – a benign exception, not a violation – in that they melt every time at exactly n degrees. (Others melt every time at exactly n − d degrees, still others fluctuate randomly in the interval, and so forth.) These are all nickel objects, indistinguishable with respect to their chemical constitution. Since we have assumed irreducible indeterminism in that world, the behaviour of any such 'miraculous' objects cannot be explained by appeal to laws holding in this world.

But we have a ready explanation! The ones which always melt at n degrees are the ones which also exist in our world. For our physical laws

[2] To be discussed in my 'Rational Belief and the Common Cause Principle', forthcoming.

pertaining to nickel are, by our present supposition, part of the essence of any nickel object existing here, and hence carried over to that other world. The scientists in that world are stumped, of course, and they might not even believe our explanation if we told them. But so it would be.

In this connection I would like to draw your attention to a report in the *Fortean Times*, Summer 1979 issue (with a reference to the *Journal of the Society for Psychical Research* 449:779). Professor John Hasted, of Birkbeck College, has recently appealed to the Everett–de Witt 'many worlds' interpretation of quantum mechanics to explain psychical phenomena.

We can leave these questions for the general theory of essences unsettled. For our purposes it will suffice that we shall always have at least the negative criterion:

> If it is physically possible for x to exist under conditions Y, but x would under those conditions not be F, then being F is not part of the essence of x.

This suffices to give us a limited but still telling access to membership of properties in the essence of an object.

3 COUNTERFACTUAL CONDITIONALS

Familiar semantic analyses of counterfactual conditionals take the form:

(3.1) $A \rightarrow B$ is true in world w exactly if the proposition $S(w, A)$ strictly implies B

We may read '$S(w,A)$' as 'A, *ceteris paribus*'. On Stalnaker's (1968) theory, or on Lewis' first theory, $S(w, A)$ is the proposition true in exactly those A-worlds which are most similar to w. This is not the correct reading for Weak Stalnaker Logic, which is my favourite, nor for the logics most easily culled from Goodman's, Chisholm's, or Sellars' initial writings on the subject; but 'A, *ceteris paribus*' fits all. And in most, or at least simple cases, it does very well to explain that the way the world would be if A were the case, everything else being equal, is just the way of the A-worlds that are most similar to ours.

But as we all recognize, similarity is a 'generic' relation, infinitely qualifiable: Peter may be similar to Paul with respect to height or hair colouring, while dissimilar with respect to education or social conscience. This confronts the general theory with two alternatives for the evaluation of counterfactuals in a world w:

(A) there is a single similarity relationship $R(w)$ to be used for the evaluation of the truth value of all counterfactuals in w, for all contexts of usage

(B) for each context c of usage, there is a similarity relationship R(w, c) to be used for the evaluation of the truth values of all counter-factuals in w.

As far as I know, no one has suggested that the similarity relationship must be the same for all worlds, nor has anyone analyzed different occurrences of the same counter-factual in the same context as involving different similarity respects.

If we think of counterfactual statements as objective, scientific, reflecting facts about the real world, we should clearly opt for (A). But a large number of problems and puzzles have steered discussants in the area in the direction of (B). Not too much of this has seen its way into the literature yet; but the basic issues emerged in the review of David Lewis' book by Kit Fine (1975), in the discussion between Ellis, Jackson and Pargetter with Lewis in the *Journal of Philosophical Logic* (1977) and in Lewis' paper (1976) on Time's arrow.

These difficulties in the theory of conditionals go some ways back, and we can discuss them in connection with an old puzzle by Lewis Carroll (1894), which Bertrand Russell (Jourdain 1918: ch. 19) saw as yielding an argument for material implication.[3] Here it is, in the words of Russell's mythical soulmate, Mr. B*rtr*nd R*ss*ll:

Allen, Brown, and Carr keep a barber's shop together; so that one of them must be in during working hours. Allen has lately had an illness of such a nature that, if Allen is out, Brown must be accompanying him. Further, if Carr is out, then if Allen is out, Brown must be in for obvious business reasons. The problem is, may Carr ever go out?

Let us denote as A, B, C the propositions that Allen is out, Brown is out, Carr is out. Then the above example appears to establish

(3.2) $C \to (A \to B)$ because, whether or not Carr is out, it is physically impossible for Allen to be out without Brown

(3.3) $C \to (A \to \bar{B})$ because if Carr is out then if Allen is out, Brown must mind the shop.

Using C.I. Lewis' strict implication or David Lewis' or Stalnaker's or even Weak Stalnaker logic we deduce

(3.4) $C \to (A \to B \ \& \ \bar{B})$
(3.5) $C \to \Diamond A$

Russell, using material implication, derives (3.4) but instead of (3.5) gets merely

(3.5*) $C \to \bar{A}$

---

[3] The analysis I give here of this puzzle I owe to Richmond Thomason.

Thus, if Carr is out, the only conclusion Russell sees as warranted is that it is false that Allen is out and 'The odd part of this conclusion', he remarks, 'is that it is the one which common sense would have drawn in that particular case.'

Russell is surely correct in this. Suppose that in fact, Carr is out. Then, is it impossible that Allen is out? That is, is there no physically possible world relative to this one in which Allen is out? There certainly is; it is one in which Carr is not out. Similarly, I am in London today; is it physically impossible for me to be in New York? Not at all; we can conceive of a physically possible world in which that is so, though it is not one in which I am in London. To disallow this would amount to accepting the inference 'A, therefore it is not possible that not A', which collapses modalities.

But we know on other grounds that material implication is not a good model for the natural language conditional. Our way out is to say that the conditionals are context dependent. In the case of (3.2) the 'because' remark after it *focusses our attention* on similarity among worlds with respect to Allen's disability. In the case of (3.3) we look instead at similarity with respect to the necessities of shopkeeping. We are inclined to accept these conditionals in different contexts.

From this point of view we can even reach Russell's conclusion. For in the logics mentioned, it is assumed that

(3.6) if A is true in w, then $S(w, A)$ is true in w and only in w; so that if A is true in w then $A \rightarrow B$ is true in w if and only if B is true in it.

From this it appears that both (3.2) and (3.3) can be correctly accepted in different contexts in the same world only if it is not the case that C and A are both true. In that case, then if C is true, A is false; and that is all. (This suggests that we can draw up a trans-contextual logic, while the usual logics are meant to apply only where the context is kept constant.)

This throws an entirely different light on the function of the word 'because' and the warranting of counterfactuals. Looking back to the semantic analysis, we see that a warrant must take the form:

(3.7) $(A \rightarrow B$ because X) is true in w exactly if $S(w, A)$ implies X, and X implies B.

But if the correct choice of $S(w, A)$ depends on context, then context also determines what propositions will go into the slot of X. They elucidate what aspect of the situation the speaker is keeping fixed in his mind, *on this occasion,* when he says *'ceteris paribus'.*

4 SINGULAR REFERENCE IN CONDITIONALS

A formal pragmatics, it now appears, will treat counterfactual condition-als as context dependent in much the way that demonstratives and indexicals are. Having seen the intimate connection between counterfac-tuals and essence, and the context-dependence of the former, we may reasonably suspect the objectivity of the latter. It has not been ruled out, however, that there will be non-trivial properties of an individual, of which it will be true to assert in any context at all that may arise in our world, that the individual would have them under any physically possible conditions. I shall try to present counter-examples to that suggestion. But it will be easily surmised that in such examples, we may slip in and out of modalities *de dicto* which would defeat any conclusion, positive or negative, about their apparent subjects.

To forestall objections along that line, I wish to remind you of the considerable evidence, provided especially by David Kaplan (1979), that demonstratives (and names) in counterfactual conditionals have rigid reference.

(1) Suppose I write with chalk on the blackboard and the personal pronoun 'I' occurs among the chalkmarks. Reichenbach (1954) suggested that 'I' could then be replaced by the indexical definite description, 'the person who produced this very token', that is to say, in this case, 'the person who produced these chalkmarks'. But it does not at all seem necessarily false to say that I might not have been the one to produce these chalkmarks. It is presumably not part of the identity of these chalkmarks to have been produced by me. This seems especially clear if we imagine a chalk writing machine which produces the marks 'I did push the button but I needn't have' when someone pushes its button.

(2) 'You' denotes the person addressed. Yet the word 'you' cannot be replaced by the description 'the person (hereby) addressed'. Speaking to Richard Healey, I say, 'I am in fact speaking to you; were I now to be addressing Colin McGinn, who would you be?' The correct answer is surely 'Healey'.

(3) I am now addressing Healey; if I had now been talking to McGinn, who would Healey have been?

Examples such as these give us an overwhelming presumption in favour of rigid reference of demonstratives and names occurring as subject terms in counterfactual conditionals. Hence I continue with the examples that are specially relevant to essence. Since the context is crucial, each statement is set in a story that provides the context of usage.

(4) I lead you into my house; my living room is unexpectedly formal, everything arranged geometrically as for an official reception. The furniture is all velvet upholstered oak, except for one chair constructed

entirely in gleaming tubular metal. 'This had to be made of an alumi-
nium–titanium alloy,' I say; 'it is Grandmother's chair and needed to be
at once strong enough to support her twenty stone and light enough to be
shifted by a feeble old lady.'

(5) Next I lead you into the dining room; all the chairs are once more in
oak but to your astonishment the table is pine. 'This table would have
been oak, of course,' I explain, 'had oak not been in such short supply
during the war, due to the shipbuilding demands of the Imperial Navy.'

(6) We leave through the French doors, which I say would have been
windows if I'd had my way, and arrive in the garden, where a magician is
giving a show. He has just apparently taken a copper from a little boy's
pocket, and we are in time to hear the child say 'You did *not* find that in
my pocket! If that had been in my pocket, it would have been a silver
coin – I am sure of it, for I know exactly what is in my pockets!'

(7) Only a few paces away, in what we now perceive to be a fairly
sizeable garden tea party, a scientist is in hot debate with a literary
dilettante brandishing the *Fortean Times*. 'I tell you', the scientist says, 'if
that thing had failed to burn yellow, it would not have been a sodium
compound; but I suspect that it did burn yellow.'

Of course, it is possible still to hold the theory that essences and laws of
nature mark objective features of the real world, and that carefully made
counterfactual assertions will faithfully reflect them. In that case, the
above examples, and the countless others that more fertile imaginations
could supply, must be explained away – and I imagine they can be.

But the pervasive context-dependence of counterfactuals suggests a
different account. In any given context of utterance, the salient and
relevant features of a thing *are its identifying features in that context and
they play the role of essence* It has those features in all the contextually
relevant worlds in which it exists. But there are no features that it has in
all worlds that may appear as relevant in some context or other.
Concentration on a limited variety of contexts focussed attention on those
situations in which the salient and relevant features of the subject
consisted exactly in its location in the classificatory scheme of some special
science. Blinders paint a pretty picture.

5 PRAGMATICS OF CONDITIONALS

In conclusion I shall take up three points related to the question whether
the context-dependence of counterfactuals raises havoc with their logic or
with ordinary patterns of reasoning. It does not.

To begin, I have now proposed a very different analysis of the word
'because' in the warranting of counterfactuals, from that ordinarily

assumed. There is a traditional metaphysical surmise that the use of 'because' points to an objective connection in nature. If so, the content of the because-clause should legislate between contrary conditionals. In my diagnosis, it has a very different function, namely to exhibit the connections which the speaker has in mind in framing his conditional. Consider the dialogue:

A. If that fuse were lit, the dynamite would explode.
B. No, because no one here is sufficiently silly to light the fuse without first disconnecting it.
A. Of course, I meant, everything else being equal, the fuse still being connected, etc.

What is remarkable here, and typical, is that A shows that he had no sense of being contradicted. Yet B has clearly asserted the contrary conditional: if that fuse were lit, the dynamite would not explode since the fuse would first have been disconnected. This can not be diagnosed as A overlooking a premise that B points out, since the law of transitivity does not hold for conditionals $(A \to B, B \to C$; hence $A \to C$ is invalid). There is a disanalogy with the case of probability where A is corrected by B drawing his attention to additional relevant evidence. In such a case, A would revise his judgment. But in the present case, A needs merely to explain what he meant, and have the comfortable feeling that both are right.

The logic of conditionals becomes applicable only when all participants to the dialogue have in mind the same criteria determining the content of the tacit *ceteris paribus* clause.

One objection to this comes from professed ignorance of conditionals. I may say: 'I am sure that either Jones will definitely be caught if he goes to the train station, or else he'll definitely escape if he goes there; it all depends on the present disposition of the security forces and so I don't know *which* is the case.' If the truth of counterfactuals depends on what the speaker has in mind, how can he ever be ignorant of their truth values?

The reason is not far to seek. In his intention, he settles the relevant criterion of similarity in intension, not in extension. Thus it is either definitely true that Peter is taller than six foot, or else definitely true that he is no taller than six foot – I say that as soon as I have settled on Peter's height as a feature to be kept fixed, and for this I need not know what his height is.

A more serious objection comes from counterfactuals about contexts of usage. Earlier attempts to give accounts of the truth values (or assertability) of conditionals as mind-dependent, like emotive theories of ethics, foundered on such examples of self-application. Suppose for example that I had maintained that it is correct for me to believe that (if A then B) exactly if B follows from my body of beliefs minimally modified to accommodate A. Then I would be stopped by an example of Richmond Thomason: 'If Sally were unfaithful to me, I would (still) believe that she was faithful (for her deception would be so clever).'

No such manoeuvre will tell against a carefully framed pragmatic account. An example of this sort would go roughly as follows. Peter argues with himself

t.   If I had stayed out late, I would have missed the film.

t + 1. The conditional I uttered at t was true then, because in that context of utterance I had state of mind X.

t + 2. If at t, I had been in different state of mind Y, then in the resulting context of utterance, a contrary conditional would have been true.

t + 3. So, if I'd been in a different state of mind at t, then if I had stayed out late, I would not have missed the film.

That conclusion is unacceptable, because whether or not he was going to miss the film had absolutely nothing to do with his state of mind at that time – only with facts like how late the film began. That his argument is invalid becomes clear when we note that it is features of the context of utterance at t + 3 which determine the *ceteris paribus* clauses for conditional antecedents inside his utterances at that time. Let that context be c. The following argument is invalid:

$S(c^1, w, A) \subseteq B$

Hence: It is true in w, as understood in c, that if the real context had been c^1, then $S(c, w, A) \subseteq B$.

The argument is but a version of the familiar fallacy: if we called a donkey's tail a leg, a donkey would have five legs.

To conclude then, it seems to me that Silverberg's Stochastic Man was, like many philosophers, somewhat hasty in his metaphysical conclusions, and that a pragmatic analysis of language may help us to be less precipitate.

REFERENCES

Carroll, L. 1894. 'A logical paradox.' *Mind* 3, 436–8.

Chisholm, R. 1955. 'Law statements and counterfactual inference.' *Analysis* 15, 92–105.

Ellis, B., Jackson, F. C. and Pargetter, R. 1977. 'An objection to possible world semantics for counterfactual logics.' *Journal of Philosophical Logic* 6, 355–8.

Fine, K. 1975. Critical notice of Lewis 1973. *Mind* 84, 451–8.

Goodman, N. 1947. 'The problem of counterfactual conditionals.' *Journal of Philosophy* 44, 113–28.

Jourdain, P. E. B. 1918. *The Philosophy of Mr. B*rtr*nd R*ss*ll*. London: Allen and Unwin.

Kaplan, D. 1979. 'On the logic of demonstratives.' *Perspectives in the Philosophy of Language,* ed. Peter A. French *et al.* Minneapolis: Minnesota University Press.

Lewis, D. 1973. *Counterfactuals.* Oxford: Blackwell.

Lewis, D. 1976. 'Counterfactual dependence and Time's arrow.' Unpublished.

Lewis, D. 1977. 'Possible world semantics for counterfactual logics: a rejoinder.' *Journal of Philosophical Logic* 6, 359–61.

Locke, J. 1690. *An Essay Concerning Human Understanding,* ed. P. H. Nidditch. Oxford: Clarendon Press, 1975.

Reichenbach, H. (1954) *Nomological Statements and Admissible Operations.* Amsterdam: North Holland. Reprinted as *Laws, Modalities and Counterfactuals.* Berkeley: University of California Press, 1976.

Sellars, W. 1958. 'Counterfactuals, dispositions and causal modalities.' *Concepts, Theories, and the Mind–Body Problem,* ed. H. Feigl *et al.* Minnesota Studies in Philosophy of Science, vol. II. Minneapolis: Minnesota University Press.

Stalnaker, R. 1968. 'A theory of conditionals.' *Studies in Logical Theory,* ed. N. Rescher. Oxford: Blackwell.

Index

Index